Explorers of the New World Time Line

Written by Ann Richmond Fisher

Illustrated by Bron Smith

Teaching & Learning Company

1204 Buchanan St., P.O. Box 10
Carthage, IL 62321-0010

This book belongs to

_____

Copyright © 2007, Teaching & Learning Company

ISBN 13: 978-1-57310-523-1

ISBN 10: 1-57310-523-6

Printing No. 987654321

Teaching & Learning Company
1204 Buchanan St., P.O. Box 10
Carthage, IL 62321-0010

The purchase of this book entitles teachers to make copies for use in their individual classrooms only. This book, or any part of it, may not be reproduced in any form for any other purposes without prior written permission from the Teaching & Learning Company. It is strictly prohibited to reproduce any part of this book for an entire school or school district, or for commercial resale. The above permission is exclusive of the cover art, which may not be reproduced.

All rights reserved. Printed in the United States of America.

Table of Contents

The Earliest Explorers: **Hannu**, **Hanno**, **Pytheas**, **Fa-hsien** or **Faxian** 7

Eric the Red, Viking explorer who colonized Greenland 9

Leif Ericson, Viking explorer in North America 11

Gudridur Thorbjarnardottir, Icelandic explorer 13

Marco Polo, Venetian traveler and author 15

Ibn Battutah, Arab traveler and author . . 18

Zheng He or **Cheng Ho**, Chinese admiral, explorer and diplomat 19

Henry the Navigator, Founder of the Portuguese empire 21

Bartolomeu Dias, First European to see the southern tip of Africa. 23

John Cabot, Italian-born navigator who explored for England 25

Christopher Columbus, Italian-Spanish navigator 26

Amerigo Vespucci, Italian explorer and namesake of America. 30

Vasco da Gama, Discoverer of route from Europe to India 32

Juan Ponce de Leon, First European in Florida. 34

Vasco Núñez de Balboa, Spanish explorer in Central America. 36

Francisco Pizarro, Spanish conqueror, explorer and governor 38

Ferdinand Magellan, Portuguese explorer who sailed around the world 40

Hernando Cortez or **Hernán Cortés**, Spanish explorer and conqueror in Mexico. 43

Jacques Cartier, Discoverer of the St. Lawrence River 45

Hernando de Soto, Spanish explorer of Florida 46

Francisco Coronado, Spanish conqueror and explorer of North America's Southwest. 48

Sir Francis Drake, English navigator and pirate 50

Samuel de Champlain, Explorer, mapmaker and founder of Quebec . . . 53

Bartholomew Gosnold, Explorer of Cape Cod and Jamestown. 55

Henry Hudson, English navigator 56

Peter Minuit, Colonizer for the Dutch in North America 58

Abel Tasman, Dutch explorer of Tasmania and New Zealand. 59

Louis Hennepin, Belgian explorer in Mississippi River Valley 60

Jacques Marquette, French explorer of the Mississippi River. 61

René-Robert Cavelier, sieur de La Salle, French explorer of the Mississippi River 63

Louis Jolliet, French-Canadian explorer of the Mississippi River 66

Vitus Jonassen Bering, Danish navigator who explored Russia 67

Daniel Boone, American pioneer and explorer 68

James Cook, British explorer of the Pacific Ocean. 71

Louis Antoine de Bougainville, First Frenchman to sail around the world . . 74

Alexander Mackenzie, Scottish explorer in Canada. 75

William Clark, Co-leader of the Lewis and Clark expedition. 76

Meriwether Lewis, Co-leader of the Lewis and Clark expedition. 79

Sacagawea or Sacajawea, Guide for Lewis and Clark 81

David Livingstone, Scottish explorer of the Nile in Africa 83

John McDouall Stuart, Scottish explorer to Australia 86

Alexandrine Pieternella Françoise Tinné, Dutch explorer of the Nile River and North Africa. 88

Henry Morton Stanley, Welsh explorer in Africa 90

Robert Edwin Peary, American explorer, first to reach the North Pole 93

Mary Henrietta Kingsley, British author and explorer in Africa 95

Robert Falcon Scott, English explorer of Antarctica 96

Roald Engelbregt Grauning Amundsen, Norwegian explorer of the South Pole 98

Roy Chapman Andrews, American explorer of the Gobi Desert. 101

Richard E. Byrd, American polar explorer 103

Sir Edmund Hillary, First to reach the summit of Mount Everest 105

Additional Explorers for Further Study. . 107

Information on Additional Explorers. . . . 109

Maps 110

Bibliography 112

Dear Teacher or Family,

What better way to teach geography and history than to study the lives of important explorers? This book contains the famous names of Marco Polo, Christopher Columbus and Ferdinand Magellan. But it also includes biographies of some lesser-known—but incredibly fascinating—explorers such as Gudridur Thorbjarnardottir (called by some the greatest female explorer of all time) and Ibn Battutah (an Arab traveler and author from Morrocco). This book does not include deep-sea, underground and space explorations. Whether the explorers considered here began their adventures in Europe, Asia, Africa, Australia or America, all of them ventured into a world that was new to them.

Exploration is a quest for the unknown. Humankind has *always* been curious about unknown people and places. But what made the people in this book actually venture out and *become* explorers? Some were motivated by greed, for in the New World they could perhaps find gold or silver. Some fervently wanted to share their religious beliefs with those who were previously unreached by their faith. Others were commissioned by their governments to explore, conquer and claim new territories for their country. And some were driven simply by *wanderlust*, the "strong longing for or impulse toward wandering."

As you read the biographies included here you will find men and women from backgrounds of all kinds. Some were born into nobility; others were poor farmers. Some were college educated; others were self-educated. The lives of these explorers ended in as many different ways as they began. Some ended up rich and famous; others were unknown and impoverished at the end of their lives. Included are names that appear often in textbooks for grades 4-8, along with additional people from different regions of the world who explored a wide variety of places.

It is probably fair to say that *all* the explorers were people of courage and perseverance. Some of their great discoveries were made accidentally, some were made militarily, but many were made after years of preparation and numerous disappointments. These explorations often brought about trade and
colonization. They sometimes brought war and disease as well.

As you read these biographies and share them with your students, please follow these pointers:

- Locate specific place names on maps or globes. Small maps are provided with most biographies. A larger more complete map will be of additional help in students' understanding. *Note: Place names are often given with both the ancient and modern locations. Spellings of both people and places vary in different sources.*

- Remind students to look at the "big picture": What else was going on in the world at the same time? How did this influence the explorer?

Please note that the information in this book has been gleaned from several different reference books and online sources. Every effort has been made to verify dates and facts in more than one source. For many of the early explorers, especially, dates and other information are unclear and differ from source to source. When the dates for an explorer are in question, a notation has been made.

Letter continued on page vi.

Many of the names of people and places have alternate spellings. We have chosen the one we felt most popular, but please forgive us if it is not the spelling used in your text. When there are two common spellings for an explorer's name, we have tried to include both. The names of places have often changed through the centuries. We have often included both the ancient name and the present-day name.

It will be helpful for your students to know the following terms before using the pages in this book:

- *campaign*—an organized course of action; often a series of military operations
- *cartographer*—mapmaker
- *commissioned*—authorized to carry out a specific task. Often these explorers were commissioned by their government to command a voyage or lead an expedition.
- *desert*—to leave, give up or run away from. On many expeditions, some would *desert* their leader.
- *expedition*—a journey, trip or voyage for a particular purpose. In this book the purpose is *exploration*!
- *mutiny*—open revolt or rebellion, often by sailors or soldiers, against their leaders
- *navigator*—a guide, a person who finds the route
- *sponsor*—someone who supports, promotes or finances an event or a person's activities

Finally, I must thank my research assistants who helped immensely by probing into reference materials in search of important, interesting and verifiable facts. They also wrote first drafts of many of these explorers. To my husband, Keith, and my daughter, Betsy, I express my extreme gratitude. This book has truly been a fascinating and challenging family project.

Now to the families, teachers and students who will use this material—enjoy your own investigations into the remarkable worlds of some remarkable explorers!

Sincerely,

Ann

Ann Richmond Fisher

The Earliest Explorers

Before the days of Leif Ericson and Christopher Columbus, there were many important explorers of great courage. Following are three that have recorded explorations in the years B.C. and a fourth man who was born in the fourth century A.D. Understandably, details of their lives including exact dates are not available, but these people and their accomplishments are still worth noting.

Hannu

The first explorer

2750 B.C.: Hannu (sometimes spelled *Hennu*), an Egyptian, made an expedition to the limits of the known world. He traveled to the region at the southeastern end of the Red Sea. At the time, it was known as the land of Punt. Today it is part of modern Ethiopia and Somalia.

Hannu returned home with great riches in wood, myrrh and precious metals. He left a record of his adventures carved in rock.

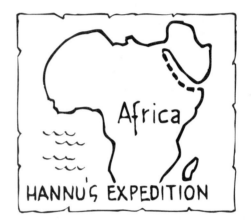

Hanno

The first explorer in western Africa

530-470 B.C.

500 B.C.: Hanno was a navigator from Carthage. About 500 B.C., he set out as the head of a large expedition to colonize Africa and start cities. It is said that he had 60 ships, each driven by 50 oars and that he started with 30,000 men and women. He took a route along the west coast of Africa. He probably sailed as far as the present-day Sierra Leone. He left some people at each place he stopped so they could begin new settlements.

When he returned to Carthage, he recorded an account of his travels on a tablet that he placed inside the temple of the Phoenician god, Moloch. The original story was written in the Phoenician language. A Greek translation exists under the title *Periplus,* which means "voyage."

Pytheas

Greek mathematician, astronomer and explorer

???-300 B.C.*

*Sources differ on exact dates in the life of this explorer.

325 B.C.: Pytheas undertook a great voyage, sailing westward beyond the Mediterranean Sea. He became the first Greek to visit Britain and the Atlantic coast of Europe. He left from his hometown of Massalía. (The city is now Marseille, France, but at that time the region was a Greek colony.) Pytheas sailed around the coast of Spain and through the Strait of Gibraltar. He had to avoid blockades put up by the Carthaginians, who were trying to monopolize all the trade in the Atlantic. Pytheas continued north along the coasts of Portugal, Spain and France. He crossed the English Channel. He continued up the west coast of Britain and landed at many places. He observed the mining and smelting of tin, the threshing of wheat and other things that were new to him.

In northern Britain, Pytheas learned of an island called Thule. It was a six-day trip to sail there. He was told it was the most northerly inhabited land, where it was daylight all the time in the summer. This could have been Iceland, but more likely it was part of Norway. It is not known if Pytheas actually sailed to Thule or not. But he did correctly describe floating discs of ice in the Arctic Sea, which would not have been known to sailors in the Mediterranean.

Fa-hsien or Faxian

Chinese Buddhist monk

???-414 A.D.*

*Sources differ on exact dates in the life of this explorer.

399 A.D.: Fa-hsien crossed Central Asia and headed to India. His goal was to visit the homeland of Buddhism. He was born in Shansi, China, although no date is given for his birth.

402: After the three-year journey, Fa-hsien arrived in northwestern India. He visited sites important to the life of Buddha. He studied extensively the early writings of his religion.

Fa-hsien traveled to Ceylon (now Sri Lanka) and continued his studies for two years.

414: Fa-hsien returned to China and translated the Buddhist writings into Chinese. The record of his travels, *Record of Buddhist Kingdoms,* contains important descriptions of India in the early A.D. 400s.

Eric the Red

Viking explorer who colonized Greenland

950-1002 A.D.*

*Sources differ on exact dates in the life of this explorer.

Note: Two Icelandic *sagas,* or long heroic tales, tell the story of the Vikings' discovery and attempted colonization of North America 500 years before Columbus sailed to the New World. The biographies of Eric the Red, Leif Ericson and Gudridur Thorbjarnardottir are based partially on these sagas.

950 A.D.: Eric was born in Jaeren, Norway. His name was Eric Thorvaldson, but he was called Eric the Red because of his red hair.

960: Eric's father was exiled from Norway for murdering a man. Eric left Norway with him. The family settled in Iceland.

980: Eric's second son, Leif Ericson, was born. Leif later became a famous Viking explorer. Eric also had two other sons and a daughter.

981-82: Eric killed two men and was forced to leave Iceland for three years. He decided to explore the land first sighted by his friend, Gunnbjörn Úlfsson, to the west of Iceland. His route took him to the island he named Greenland.

985: Eric's banishment from Iceland was over. He returned there and recruited people to sail with him to the new land he had discovered. Although the island was covered in ice, he called it Greenland to make it sound nicer and to encourage settlers to go with him.

986: Eric sailed for Greenland with 25 ships and approximately 400-500 people who wanted to form colonies in Greenland. Only 14 of the vessels and 350 people completed the journey.

When they arrived, the eastern coast of Greenland was covered in ice, so the colonists rounded Cape Farewell in the south. They founded two settlements, Brattahild (near what is now Julianehab) and Godthab (or Nuuk). Both communities were on the western coast. The settlers farmed the land and raised cattle, hogs and sheep. They hunted bears, caribou and other animals. They fished, as they had before in Iceland.

After doing well for awhile, the settlements experienced unusually cold weather. That prompted some of the settlers to return to Iceland. In time, all of the settlers disappeared. It is unclear if they were attacked by Inuit people or if they died from illnesses and starvation. After the 13th century these settlements disappeared. However, Eric the Red's exploration opened the door to centuries of explorations of the area. Later, other northern Europeans also attempted to make colonies in Greenland.

It is reported that Eric planned to lead an expedition west of Greenland in search of more land. On the way to his ship, however, he fell from his horse. He thought that was a sign of trouble ahead and so he refused to make the journey.

1001: Eric's son, Leif, set off on a voyage that eventually led him to North America. Eric died the winter after Leif returned home.

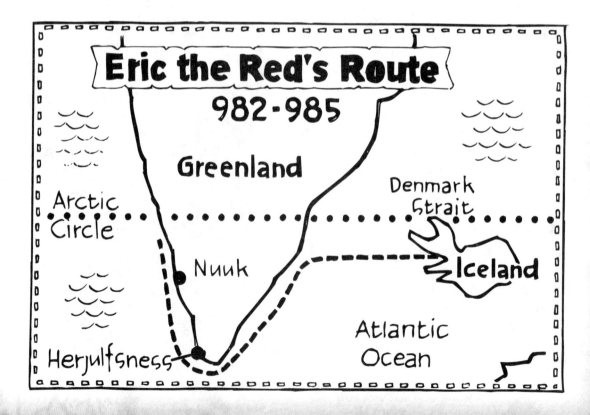

Leif Ericson

Viking explorer in North America

980-1020*

*Sources differ on exact dates in the life of this explorer.

Note: Two Icelandic *sagas,* or long heroic tales, tell the story of the Vikings' discovery and attempted colonization of North America 500 years before Columbus sailed to the New World. The biographies of Eric the Red, Leif Ericson and Gudridur Thorbjarnardottir are based partially on these sagas.

980: Leif Ericson was born in Iceland. His name is also sometimes spelled *Ericsson*, *Eiríksson* or *Erikson*. He was the second son of Eric the Red.

982: Leif's father, Eric the Red, moved the family to Greenland, where he established two colonies. Leif grew up in Brattahild, now the Eskimo village of Kagsiarsuk, on the southwest coast of Greenland.

985: According to some sources, an Icelandic trader named Bjarni Herjólfsson was blown off course and sighted a new land far to the west and south of Greenland. His story caught the attention of Eric the Red, Leif's father. Later, Eric sent Leif to explore the land.

999: At age 19, Leif voyaged to Norway. He spent a winter at the court of Norway's Christian king, Olav I Tryggvason. According to some reports, when Leif returned to Greenland, he converted his mother to Christianity. She built the first church in Greenland.

1001: Some sagas say that Leif bought Bjarni Herjólfsson's ship and set sail with one ship and 35 men. He followed Bjarni's course in reverse. First Leif and his crew put ashore at a place described as a barren land of flat rocks backed by huge ice mountains. Then they went to sea again and arrived at a level wooded land with broad stretches of sand. They called this Markland, or Forest Land. Once again, they sailed southward. This time they found a land that was green with wheat, trees and wild grapes. They named it Vinland, or Wineland. Here they built shelters and spent the winter. In fact, some reports say that Leif built a small village and spent the winter making a cargo of timber for his return voyage to Greenland.

For many years, the location of Leif's "Vinland" was debated. In 1963, a Norwegian archaeological expedition found remains of a Viking settlement on the Strait of Belle Isle and were able to date the remains to about A.D.1000. Because of this, most historians believe he landed in Canada, on the northern tip of Newfoundland.

On his return journey, Leif encountered a wrecked trading vessel and rescued its crew of 15. For this deed he received the entire rich cargo and the nickname "Leif the Lucky."

Once back in Greenland, Leif gave glowing accounts of the land he had discovered. He told of its riches and thick, tall forests. He described rushing rivers full of salmon and grassy meadows that would be plentiful pasture for livestock.

Thorfinn Karlsefni, an Icelander, established a colony in Vinland. The colony of 160 men and five women spent three years at Leif's wintering place. Karsefni's son Snorri was the first child of European descent born in North America.

1025: "Leif the Lucky" probably died around the year 1025, at the age of 45.

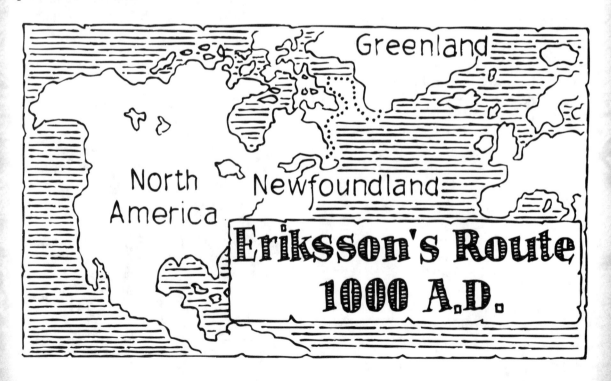

Gudridur Thorbjarnardottir

Icelandic explorer

Lived about the year 1000.*

*Many dates in the life of this explorer are unknown.

Note: Two Icelandic *sagas,* or long heroic tales, tell the story of the Vikings' discovery and attempted colonization of North America 500 years before Columbus sailed to the New World. The biographies of Eric the Red, Leif Ericson and Gudridur Thorbjarnardottir are based partially on these sagas.

The late 900s*: Gudridur Thorbjarnardottir was born in Iceland. Her parents were Thorbjarner Vífilsson and Hallveig Einarsdóttir. Her paternal grandfather had come to Iceland as a slave but later gained his freedom.

Gudridur was also the foster-daughter of Orm and Halldís of Arnastapi, Iceland. While staying with her foster family, Gudridur fell in love with a young man. He was the son of a slave who had become very successful in his own life. He wanted to marry Gudridur, but her foster father would not allow her to marry a slave's son. Instead, her foster father decided to move the family to Greenland.

996: About 10 years after Eric the Red led a settlement group to Greenland, Gudridur's family set sail for the new land. Many people traveled with them on the difficult voyage in terrible weather. Gudridur's foster parents, and many others, died along the way.

Illustration of a sculpture by Ásmundur Sveinsson, depicting Gudridur, dating from 1939.

Because Eric the Red was a friend of Thorbjarnar (Gudridur's birth father), he gave Gudridur some land near his own. Eventually Gudridur met and married Eric's oldest son, Thorsteinn. Legends say that Gudridur heard the stories of her brother-in-law, Leif Ericson, and she longed to see the New World he had just discovered. Gudridur and Thorsteinn moved to Vinland after Thorsteinn's brother Thorvaldur had been killed there by the natives. It was another long, difficult voyage with much sickness. This time, Thorsteinn died, leaving Gudridur a widow. Apparently, she returned to Greenland.

After time, Gudridur met and married Thorfinnur Karlsefni, a wealthy merchant who came to Greenland from Iceland. He was of royal descent and was the best sailor of his day. He also wanted to explore the New World. Thorfinnur and Gudridur led an expedition to explore and settle in Vinland. According to the *Saga of Greenlanders,* there were 60 men and five women on their ship. The settlers also took livestock.

Leif Ericson gave Thorfinnur permission to use the houses Leif had built in Vinland during his expedition there.

1004: Gudridur became the first European woman to give birth on the American continent. Gudridur and Thorfinnur named their son Snorri Thorfinsson. The site of his birth is thought to be in Newfoundland at l'Anse aux Meadows. It was discovered more than 40 years ago by archaeologists and is now a United Nations World Heritage Site.

1004-1007: Scholars believe Gudridur's group explored Vinland as far south as Manhattan. But tensions grew between the Europeans and the natives. So Gudridur's group planned to return to Greenland. Before they left, natives killed her husband. She was a widow for the second time.

1007: Gudridur returned to her native Iceland. She converted to Christianity and immersed herself completely in this new religion. She eventually became a nun.

Gudridur undertook a pilgrimage and traveled alone to visit the Pope in Rome. She crossed Europe by foot. She gave the Pope a report on Christianity in both Iceland and Greenland.

The date of Gudridur's death is unknown.

The president of Iceland has call Gudridur "the greatest female explorer of all time."

Up the hill from a recently uncovered turf house that could have been Gudridur's, archaeologists found another site with a church dated to the year 1030. A statue of Gudridur and her son stands in the cemetery. Both mother and son are facing Newfoundland.

Marco Polo

Venetian traveler and author

1254-1323*

*Sources differ on exact dates in the life of this explorer.

1254: Marco Polo was born in Venice into a merchant family. Venice was one of the most important trade centers in Europe at that time. Growing up in Venice, young Marco would have met many traders from China, Rome and other places.

1260: Marco's writings say that in 1260, his father Nicolo and his uncle Maffeo Polo, who were jewel merchants, left Venice. They traveled to ports on the Black Sea, including Constantinople (now Istanbul, Turkey) and Soldaia (now Sudak, Ukraine). The three travelers continued farther east to trading cities on the Volga River in Russia.

1262: A war broke out behind Marco's father and uncle, preventing them from turning around and going back home. So they traveled even farther east to Bukhoro (in present-day Uzbekistan).

1265: After three years in Bukhoro, the Polos joined a diplomatic mission going to the court of Kublai Khan. He was the Mongol emperor of China. The Khan was warm and friendly. He told the Polos he wanted to learn more about Christianity. He asked the Polos to return to Europe and ask the Pope to send 100 Christian scholars who could explain that religion to him.

1269: Nicolo and Maffeo returned to Venice after being away for many years, in order to satisfy the Khan's request. The Pope appointed just two missionaries to go with them on their trip back to Kublai Khan's court.

1271: Nicolo and Maffeo headed back to China. They took Marco and the missionaries with them. They traveled from Venice to Palestine. But the missionaries were worried about hazardous conditions along the route and they abandoned the trip. The three Polos continued the journey. They probably traveled overland through Persia (now Iran) to Hormuz at the mouth of the Persian Gulf, then north through Persia and Afghanistan. They crossed long, dry stretches of the Gobi Desert.

1275: Early in the year, Marco Polo, his father and his uncle arrived at Kublai Khan's summer court in Peking. Marco was 21 years old at the time.

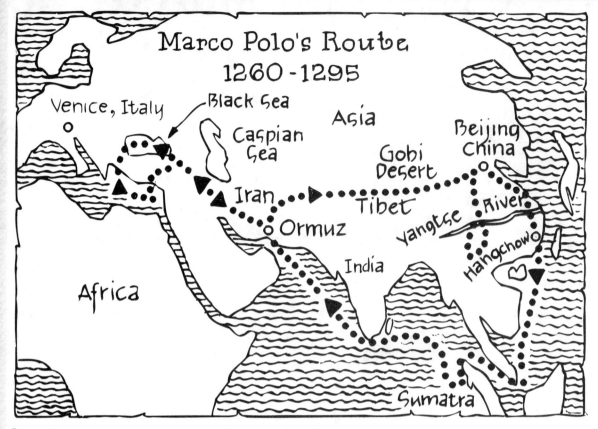

1275-1278: Marco became a favorite of Kublai Khan. Marco was a great storyteller and an interesting conversationalist. The Khan sent him on many diplomatic missions throughout his empire. Marco came back with new stories and observations about the lands he had visited. Marco reported that Kublai Khan also made him governor for three years of the large commercial city of Yangzhou. Some scholars today doubt this claim.

The Polos became wealthy in China. But they began to be afraid that jealous men in the Kublai Khan's court would hurt them when the Khan died. They asked permission to return to Venice many times, but Kublai Khan refused. Then an envoy arrived from the Khan of Persia. He asked Kublai Khan for a young Mongol princess for a bride. The Polos said the princess's journey should be guarded by men like themselves.

1292: Finally, the Khan allowed the Polos to return to Venice, escorting the princess to Iran on the way. Because there was danger that the group could meet robbers and enemies of the Khan along the overland trade routes, a great fleet of ships was built so they could travel by sea. In addition to the three Polos and the princess, there were also 600 Chinese noblemen on the voyage.

The group left the southern Chinese port city of Quanzhou (in the current Fujian Province) and sailed to what is now Sri Lanka. Then they went to India and the Persian Gulf. They touched the East African coast. The three Polos and the princess were safe, but the voyage was hazardous. Only 18 of the 600 noblemen lived to reach Persia (Iran). After the princess was delivered safely to Iran, the three men traveled through Turkey, took a ship to Constantinople and then finally traveled to Venice. They arrived home in 1295. They brought with them the great wealth that they had acquired in China.

1298: Marco became involved in a naval conflict between Venice and the rival city of Genoa. He was captured, along with 7000 others, when the Genoese navy defeated his Venetian fleet. During his one-year imprisonment, he passed the time by telling stories. Marco's stories caught the attention of a writer from Pisa named Rustichello. Rustichello then prepared a book of Marco's travels in a French dialect. Scholars disagree as to whether Rustichello wrote the book himself or simply helped Marco to write it. It is thought that Rustichello sometimes embellished Marco's stories. Nevertheless, the book was very popular. Soon it was translated into more languages. The English translation of the original title of the book was *The Description of the World.*

1299: Marco was released from prison. He returned to Venice and engaged in trade. He married and had three daughters. His name appears in the court records of his time in many lawsuits regarding property and money.

1323: Marco Polo died in Venice.

Marco Polo's description of his travels greatly influenced European readers. Mapmakers referred to it for information on Asia. Merchants were inspired by it when they planned trading trips. Mariners studied it when they wanted to find a water route to India in the 15th century. Christopher Columbus owned a Latin translation of Marco's book, which he read carefully. His copy still exists, along with his handwritten notes in the margins. Columbus relied heavily on Marco Polo's geography when planning his own voyage to reach Asian markets by sailing west from Europe.

Ibn Battutah

Arab traveler and author

1304-1368*

*Sources differ on exact dates in the life of this explorer.

1304: Ibn Battutah was born in Tangier, Morocco, on February 24. His full name was Muhammad ibn Abdullah ibn Batuta. (Sources list both the spellings *Battutah* and *Batuta*.) He was a Muslim.

1325: Ibn Battutah took his first journey. It was a religious pilgrimage to Mecca, Saudia Arabia. During this first trip he became enthusiastic about traveling. He vowed to visit as many parts of the world as possible. Ibn Battutah made this rule for himself: "Never travel any road a second time." He wanted to learn about new countries and new peoples. As he became more and more famous, he made a living from his travels. Many rulers, sultans and high officials gave generously to him so that he could continue his journeys.

From 1325 to 1354, he traveled through North Africa, the Middle East, East Africa, Central Asia, India, Southeast Asia and possibly even China. He covered perhaps 75,000 miles.

Ibn Battutah wrote about his wanderings in a book called *Rihlah*. The title means "travels." It is an important source for the history and geography of the medieval Muslim world. The book includes descriptions of the count of Constantinople (presently Istanbul) and the Black Death of Baghdad in 1348. Ibn Battutah finished his book in December 1357.

1368: Ibn Battutah died in a town in Morocco where he held the office of judge. He was buried in his native Tangier.

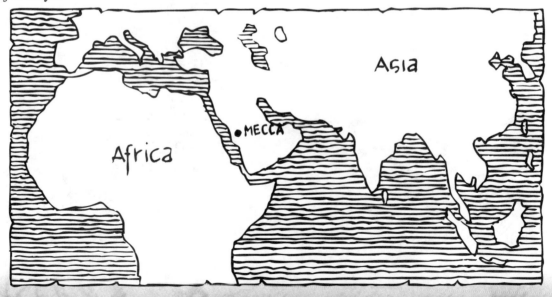

Zheng He or Cheng Ho

Chinese admiral, explorer and diplomat

1371-1433*

*Sources differ on exact dates in the life of this explorer.

1371: Zheng He (Cheng Ho) was born in China in Kunyang, in China's southern Yunnan province. At birth he was named Ma Sanpao, or *Three Jewels*. He was born into a Muslim family.

1382: At age 11, Zheng was seized and sent into the army. The army was under the command of Chu Ti, the Prince of Yen.

1402: The Prince of Yen took the throne away from his 16-year-old nephew with the help of Zheng. The prince became the Emperor Ming Yonglo.

1404: To thank Zheng for his help, the emperor named Zheng the Grand Imperial Eunuch and changed his family name from Ma to Zheng. He became a close aide of Emperor Ming Yonglo.

1405: Zheng began his first of seven sea explorations. He had 62 large ships and a 27,800-man crew. He visited much of southern Asia, including Indochina, Indonesia, Malacca, Ceylon (Sri Lanka) and Calicut (in India).

1407-1409: On his second expedition, Zheng sailed westward through the Indian Ocean to Calicut.

1409-1421: Zheng took four more voyages during this time period. The ships with which he sailed were 400 feet high, weighed 700 tons, had multiple decks with 50 or 60 cabins, and carried several hundred people.

On these journeys he visited Southeast Asia, India, Ceylon, the Persian Gulf, East Africa and Egypt. He returned with shiploads of exotic goods. He brought back diplomats from more than 30 countries who wished to pay their respects to Emperor Yonglo.

1431: Zheng took his seventh and last trip. He went to many of the same places he had been previously, including Southeast Asia and ports of the Indian Ocean.

1433: While on his seventh trip he reached the island of Hormuz in the Persian Gulf.

Zheng died in Calicut while returning to China. The legend claims his body was not returned to China for burial. (The exact year of his death is uncertain, but historical reports show it was between 1433 and 1436.)

After his death, China returned to its isolationism, and most of Zheng's carefully drawn sailing charts were destroyed. His voyages exceeded any other expedition that had taken place at that time.

On his voyages, Zheng and his crewmates traveled as far west as eastern Africa and as far south as Java and Sumatra. Zheng's explorations led to trade with those same countries. The explorer began diplomatic relations with at least 35 countries in the South Pacific, the Indian Ocean, the Red Sea and the Persian Gulf.

His exploration encouraged Chinese to move to other countries. They were able to influence many countries with the Chinese culture because of his authority.

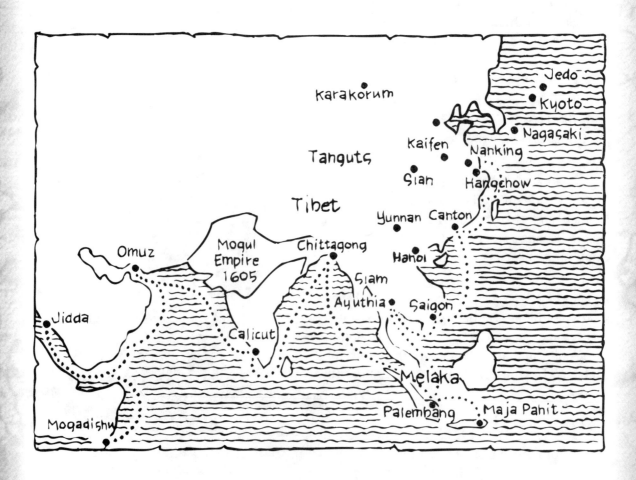

Henry the Navigator

Founder of the Portuguese empire

1394-1460

1394: Prince Henry was born in Orporto, Portugal, on March 4, 1394. Henry was the third surviving son of King John I of Portugal. His mother, Philippa, was the daughter of the English Duke of John of Gaunt.

1415: Prince Henry played a leading role in the successful Portuguese attack on Ceuta in Morocco, on the northern coast of Africa. While Henry was there, he became interested in the continent of Africa. He decided to send out expeditions to explore the African coasts.

Then Prince Henry set up a base not far from the port of Lagos in the southwestern tip of Portugal. On a peninsula called Sagres, he established an observatory and the first school for navigators in Europe. The most brilliant scholars did pioneer work there in navigation, astronomy and mapmaking. Prince Henry made improvements in the art of shipbuilding. A new sailing ship, the caravel, was designed at Sagres.

1418: Prince Henry began sponsoring several voyages of exploration. While he made no voyages himself, his Portuguese pupils explored the African coast and many nearby islands. He sent out most of his expeditions from Sagres.

1419: Henry was made governor of a region in Portugal.

1420: Henry's navigators reached Madeira. In this same year, he was appointed grand master of the Order of Christ, which was a wealthy Portuguese crusading order. This group helped to finance Henry's early expeditions.

1427: One of Henry's navigators discovered the Azores, which were settled by the Portuguese in 1439.

1434: Henry's navigator, Gil Eanes, sailed around Cape Bojadore, along the coast of present-day Morocco. It was seen as Henry's first important success. Henry was granted a monopoly on trade and conquest beyond Cape Bojadore in 1443.

1445: Henry's navigators rounded Cape Vert (along the coast of Senegal) in 1445. They reached the mouth of the Gambia River about 1446.

1460: Prince Henry died in Sagres on November 13. Because he had spent his entire fortune on his projects and expeditions, Henry was in debt when he died.

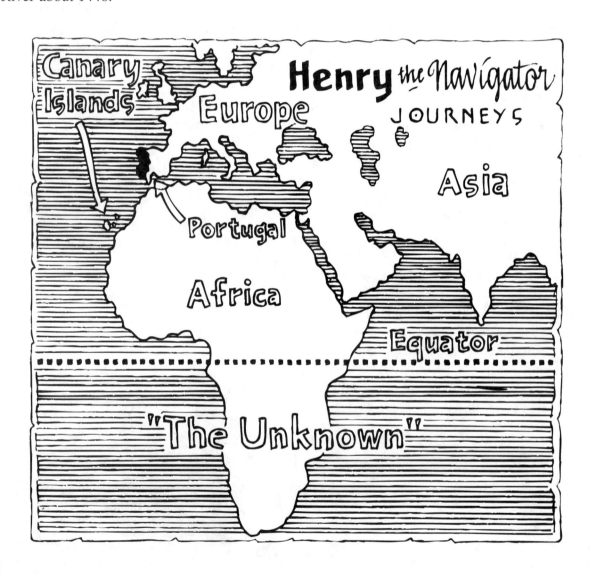

Bartolomeu Dias

First European to see the southern tip of Africa

1450-1500*

*Sources differ on exact dates in the life of this explorer.

1450: Bartolomeu Dias (also spelled *Diaz*) was born in Portugal, during the height of Italy's control of the Mediterranean Sea. As a young man, Dias became involved with the dangerous African gold and ivory trade. He was successful enough to become a captain. At that time, Italian cities were becoming rich from their trade with India and other parts of Asia. Portugal was eager to share in this trade, too. But the Italians controlled the Mediterranean Sea, which was the main route to Asia. The groundwork had been laid by Henry the Navigator. Exploration continued under Henry's nephew, King John II.

1486: Diogo Cam, a Portuguese explorer, sent King John II word that he had sailed past the mouth of the Congo River, near present-day Walvis Bay. The king chose Dias to lead a new expedition to sail to the southern tip of Africa to find a new trade route with Asia.

1487: In August, Dias sailed with three ships from Lisbon, Portugal, to the Walvis Bay area in Africa. He continued sailing southward and kept close to the coast of Africa until he reached present-day Luderitz, Namibia.

1488: Dias continued his southward journey. When a gale blew his fleet past the southern tip, or cape, of Africa, he continued to sail east. When he didn't find any land, he turned north. He had, unknowingly, rounded the tip of Africa. Just before Dias entered the Indian Ocean, his crew forced him to turn around. On the return voyage, Dias spotted the tip of Africa and named it *Cabo Tormentoso,* or Stormy Cape.

Dias had just opened a sea route from Europe to the Far East. European traders and politicians felt this was essential to the prosperity of Europe.

In December 1488, Dias returned to Lisbon, and King John of Portugal renamed the cape the Cape of Good Hope.

1497: Dias assisted in building the ships for Vasco da Gama's voyage. Dias traveled under da Gama as far as the Cape Verde Islands to help him navigate the voyage.

Dias remained on the islands to command Fort Mina, where he traded gold and slaves. Da Gama reached India on May 20, 1498.

1500: Dias was a captain of a ship in a fleet commanded by Pedro Alvares Cabral. The goal of the journey was to reach India again. But the fleet headed west in the South Atlantic Ocean and accidentally made the first discovery of Brazil. After meeting the natives of Brazil, Cabral's fleet set back across the South Atlantic. On May 24, the fleet encountered a cyclone, and four ships, including the one carrying Dias, sank. Dias and the rest of his men drowned. Cabral finished the voyage with his six remaining ships and established Portugal's trade route with India.

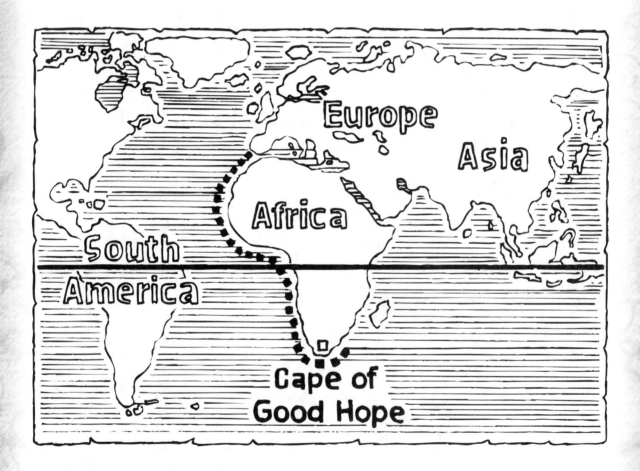

John Cabot

Italian-born navigator
who explored for England

1450-1499*

*Sources differ on exact dates in the life of this explorer.

1450: Cabot was born in Genoa, Italy.

1476: He became a naturalized citizen of Venice, Italy.

1484: Cabot moved to England and sailed under the English flag.

1493: People in England learned that Christopher Columbus had made the westward passage to Asia. Cabot and his supporters began to make plans for a more direct route.

1496: Cabot's expedition was authorized by King Henry VII. This same king did not help Christopher Columbus when he wanted support for his westward journey five years earlier.

1497: Cabot sailed from Bristol, England, on the *Matthew*. He sailed northwesterly until he came to the coast of North America, but Cabot thought he had reached northeastern Asia. He probably discovered Cape Breton Island, then sailed along the Labrador, Newfoundland and New England coasts. This made John Cabot the first European to reach the North American mainland since the Vikings. England later claimed all of North America because of Cabot reaching the mainland.

When he returned to England, Cabot was welcomed by the people and granted an annual pension by Henry VII.

1498: Cabot left Bristol on his second voyage. He planned to return to the coastline he had already found and sail southward. Then he could find Japan, he thought. His supplies ran low as he reached the coast of Greenland. His crews mutinied because of harsh weather conditions. No clear records have survived, but it is thought he was forced to return to England.

1499: It is thought that Cabot died in Bristol after he returned there around 1499, although some reports say that Cabot was lost at sea in 1498.

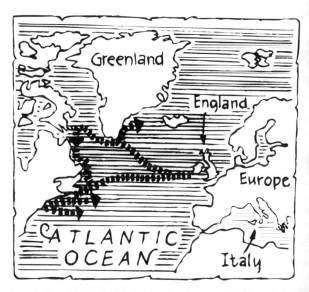

Christopher Columbus

Italian-Spanish navigator

1451-1506

1451: Christopher Columbus was born in the port city of Genoa, Italy. (In Italian, his name was Cristofro Columbo.) His father, Domenico Columbus, was a weaver of wool. Christopher helped in the family business. He had little or no schooling, but he was quick to learn from experience.

1473: He worked for his father until he was 22. He may have gone out in the fishing boats, and he probably did some business for his father by boat. He made at least one trip to the North African coast.

1476: Columbus sailed as a common seaman on a merchant ship. While sailing near Portugal, the ship was attacked and went down. Columbus swam ashore to Lisbon. He settled in Lisbon where his brother, Bartholomew, worked as a mapmaker.

Portugal was the world's leader in sailing expeditions, and Columbus saw his chance to become a captain. He learned to speak and read Portuguese and Castilian (language of Spain). He became a chart maker to earn a living.

1479: Columbus married Dona Felipa Perestrello, daughter of the governor of the island of Porto Santo. She was of high social rank, which allowed Columbus to meet important officials. Dona Felipa gave Columbus her father's charts and documents.

1480: Columbus and Dona's only son from this marriage, Diego, was born. Dona Felipa died soon after the birth of Diego.

1481: Columbus entered the service of King John II of Portugal and sailed to the Gold Coast of Africa.

After studying maps, charts, the writings of Marco Polo and others, Columbus decided the world was 25% smaller than was previously thought. (The educated men of his time knew the world was round and that Asia was west of Europe, but no one knew how far.) As a result, he wanted to sail west to reach Asia because he thought it would be a quicker route.

1484: Columbus submitted a proposal to King John II of Portugal to sail across the Atlantic Ocean. The king's committee rejected the plan because they thought it was unsound.

1485: Columbus went to Spain to try to convince the rulers there to fund his expedition.

1486: Through a series of meetings with influential persons, he was introduced to Queen Isabella of Spain. He spent seven years trying to gain support from Isabella I for his plans.

Around this same time, Columbus took Beatriz Enriquez de Harana as his common-law wife. She gave birth to their son, Ferdinand Columbus.

1492: King Ferdinand and Queen Isabella of Spain finally agreed to sponsor the plan by Columbus to sail west. On August 3, Columbus set sail from the harbor town of Palos with about 90 men, across the Atlantic Ocean. The *Santa Maria* was contracted for the journey. Because the town of Palos had offended the Spanish rulers, they were ordered to furnish two additional ships, the *Niña* and the *Pinta*, as a penalty. Columbus lead the *Santa Maria*, and two captains from Palos directed the other two. The *Santa Maria* was about 100 feet long. The *Niña* and the *Pinta* were each about 50 feet long.

Columbus kept two logs of the journey. In one log, he recorded the distance he thought they had actually traveled. In the other log, he wrote a shorter estimate that he showed to the crews, so they wouldn't fear the long distance from home. The shorter record was closer to the actual mileage than the distance Columbus thought.

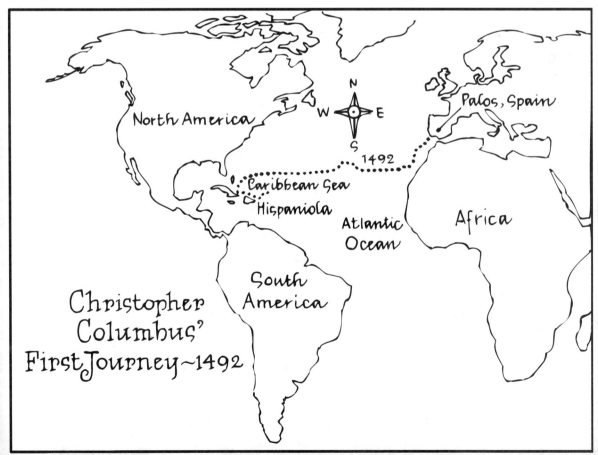

Christopher Columbus' First Journey ~ 1492

Sailing was mostly smooth but the crews began grumbling about the inability of Columbus to find his way. They knew the compass was off from true north. But on October 12, they spotted land. They landed on Guanahani, as named by the natives, in the Bahamas. Columbus renamed it San Salvador. The trip had taken 70 days.

Columbus and his men were greeted by a group of islanders who were friendly but curious. Columbus took possession of the island while the natives watched. At once, the Spaniards began trading with the natives.

Columbus was looking for a route to Asia. What he found was the New World, the Americas. To his dying day, Columbus thought he had reached Asia.

In the next few weeks they discovered other islands that today are known as Rum Cay, Long Island, Crooked Island, Hispaniola (the Dominican Republic and Haiti) and Cuba. The *Santa Maria* was wrecked off the coast of Hispaniola. From its timber, Columbus built a small fort at La Navidad. He selected 39 sailors to stay as colonists.

1493: The *Niña* and the *Pinta* began their journey home in January. Columbus and his crew arrived back in Spain in March. Columbus brought back gold, cloth, ornaments, exotic plants and animals, and even several natives. In Barcelona, the Spanish king and queen gave Columbus many titles and honors.

In September, Columbus started on his second voyage with 17 ships and 1500 men. This time the trip took 39 days. He landed on the islands of Dominica, Guadeloupe and Antigua. Columbus also went to Puerto Rico, which is the closest he came to landing on soil that would later become part of the United States. This is the basis of the claim that Columbus "discovered America."

In November, Columbus visited La Navidad and found the fort had been destroyed and his men had been killed. He started the new colony of Isabella, which was the first European settlement in the new world.

Santa Maria

1494: Columbus explored the coasts of Jamaica and Cuba. He returned to Isabella to govern the colony but was not a good leader. Columbus established a new capital, Santo Domingo. He left his brother, Bartholomew, in charge as the governor.

1496: Columbus returned to Spain.

1498: Columbus left on his third voyage. He landed on Trinidad. He discovered several more small islands before traveling back to Santo Domingo. He found the colony in revolt against his brother.

1499: The colony of Santo Domingo had sent complaints to the king of Spain. The king sent a replacement to rule Santo Domingo.

1500: Bobadilla, the replacement for Columbus, arrived at Santo Domingo. He had Columbus and his brother arrested and shipped back to Spain in chains. Columbus refused to have the chains removed until the queen saw him. The monarchs released the brothers and restored the titles formerly held by Columbus. They did not give him back his post in Santo Domingo.

1502: Columbus began his final voyage. He was still looking for a passage to Asia. He was given four ships in poor condition. He lost two of the ships while exploring Central America. The remaining two ships ran aground in Jamaica. He sent for help from Hispanola, but it took almost a year for the help to arrive.

1504: Columbus and his crew were taken back to Spain. Columbus would not sail again. In his last few months of life, Columbus battled illness. He also tried, but failed, to have his full privileges restored by the king. Nevertheless, he was quite wealthy because of the gold he had brought back from Hispaniola.

1506: Columbus died at Valladolid, Spain.

1533: His remains were transferred to a monastery in Seville.

1542: His body was transferred to Santo Domingo.

1795: A box believed to have the bones of Columbus was moved to Havana, Cuba.

1877: Another casket bearing the name of Columbus was found in Santo Domingo.

1899: The first set of bones that was taken to Havana was moved again to Seville.

Columbus was not the first explorer to cross the Atlantic—the Vikings crossed the ocean in 1000 AD. Christopher Columbus did, however, open up new trade routes to the west that had not previously been known. Christopher Columbus will always be remembered as a great explorer who found his way to the New World again and again.

Amerigo Vespucci

Italian explorer and namesake of America

1454-1512*

*Sources differ on exact dates in the life of this explorer.

1454: Amerigo Vespucci was born in Florence, Italy. He was first educated by his uncle, a noted philosopher. Vespucci then taught himself much about science. His father was a notary.

1483: After his father died, Vespucci went to work for the Medici family. The Medicis were merchants who outfitted many of the Italian explorers headed for the New World, including Christopher Columbus on his third voyage.

1491: Vespucci moved to Seville, Spain, to work in the trading business. He was in Spain when Columbus returned from his first voyage to the New World. He helped Columbus get ships ready for his second and third journeys to the west.

1497: After watching the success of Columbus, Vespucci asked the King of Castille, Ferdinand, for his support for a voyage. He was given three ships and left from Cadiz, Spain. Vespucci claimed to have reached the coast of South America, probably near Guiana. He may have sailed as far as the Gulf of St. Lawrence, before returning to Cadiz on October 15, 1498. Many scholars argue that this voyage never even took place. But if Vespucci's claims were accurate, he was the first European to land on the American continent, for at that time Columbus had only reached the outlying islands.

1499: Vespucci left from Cadiz for his second voyage. On this journey, he discovered the Amazon River and made many scientific observations of the constellations and sea currents before returning to Spain in September of 1500. After this journey, Vespucci became very ill, but he recovered enough to continue plans for further expeditions.

1501: Vespucci was supported by Portugal rather than Spain. He sailed along the coast of South America, as far south as the Rio de la Plata, in modern-day Argentina.

1503 and 1505: Vespucci probably made two more expeditions. It is believed that he established an agency at the Tropic of Capricorn. He probably explored South America and collected gold from there.

1505: Vespucci married Maria Cerezo. She died between 1523 and 1524, and left no children.

1507: A German amateur mapmaker and minister, Martin Waldseemuller, made a map of the New World and decided to name it after Amerigo Vespucci. The name is now known to us as *America*. When another continent north of South America was found, it was also named for him.

1508: Vespucci was granted the title *piloto mayor de España*, making him the ceremonial chief navigator of all Spanish voyages.

1512: Vespucci died in Seville, Spain.

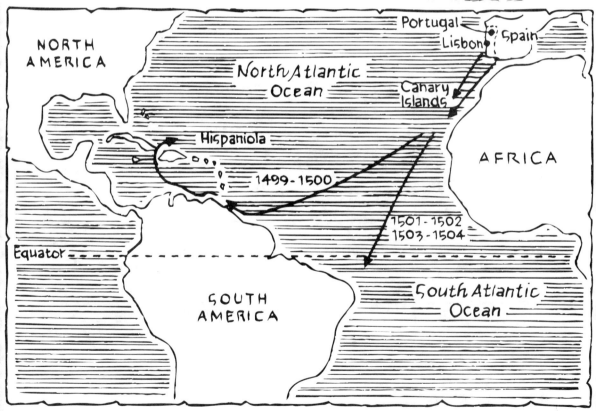

Vasco da Gama

Discoverer of route from Europe to India

1460-1524*

*Sources differ on exact dates in the life of this explorer.

1460*: Vasco da Gama was born in Sines, Portugal, on the southwest coast of the Iberian Peninsula. His father commanded the Portuguese fortress there. As a young man, da Gama entered the service of King John II of Portugal and helped defend Portugal's colonies against French attacks.

1497: Vasco da Gama was appointed to sail around the Cape of Good Hope and reach India. (The Cape had already been discovered by Bartolomeu Dias.) His father had been appointed to lead this expedition but died before he could carry it out. Da Gama was given four vessels, and he made his brother Paolo captain of one of them. They left from Lisbon, Portugal, on July 8. Dias accompanied the fleet on the first part of the voyage, and in November, da Gama rounded the southern tip of Africa.

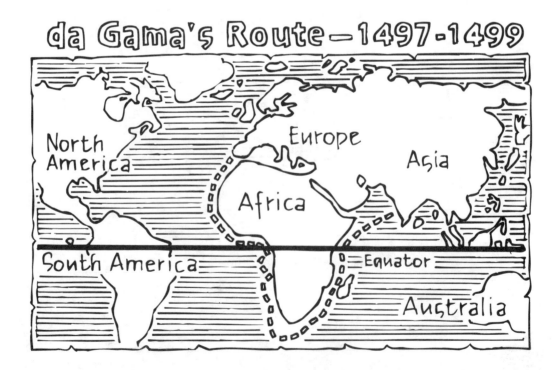

1498: Vasco da Gama and his crew landed at Calicut, on the southwestern coast of India, on May 20. He worked to earn the respect of the Hindu ruler of Calicut. This ruler was influenced by Muslim traders who felt threatened by the arrival of Europeans. The Muslim traders also wanted to trade with India and didn't like the competition of the Europeans.

1499: Da Gama returned to Portugal in 1499 with spices and gold, but only two of his original ships. Of his original 170 crew members, only 55 remained. The rest died on the journey due to *scurvy*, a disease caused by a lack of vitamin C. Da Gama's brother, Paolo, was among those who died. Upon his return, da Gama was granted the title *dom*, and given the title Admiral of the Indian Ocean.

1502: Pedro Alvares Cabral was sent by Portugal on a new expedition to India, but many of his men were killed in India. King Manuel of Portugal ordered Vasco da Gama to make another trip to India with 20 ships. On this trip, he killed hundreds of Muslim traders and burned many of their ships, but he was able to sign several treaties with Indian rulers and returned with several tons of valuable goods.

1503: Da Gama returned to Portugal, where he was once again given many honors and awards by the king. Da Gama was married and raised a family while serving as an advisor to the king.

1519: Vasco da Gama was given the title of Count of Vidigueira, a large town in Portugal.

1524: Under King John III of Portugal, da Gama was made Viceroy to India. He was the representative of the king in Portugal's colonies there. The king wanted him to curb corruption by Portuguese officials living in India. Vasco da Gama died three months later in Cochin, India, on December 24. In 1539 his remains were returned to Portugal.

Vasco da Gama's voyages brought great wealth to Portugal. As a result of his exploration, his country became one of the greatest powers in Europe because it controlled the route to the East Indies.

Juan Ponce de Leon

First European in Florida

1460-1524*

*Sources differ on exact dates in the life of this explorer.

1460: Juan Ponce de Leon was born in Santervas, Spain. As a boy, he worked in the royal court delivering messages.

Early 1490s: Ponce de Leon fought at Granada, in southern Spain, against the Moors.

1493: Ponce de Leon went with Christopher Columbus on his second journey to the New World. They went to the island of Hispaniola. (Today that island is made up of Haiti and the Dominican Republic.) Instead of returning to Spain with Columbus, Ponce de Leon stayed on the island to protect the Spanish colony. He fought to put down several rebellions by the natives of the island.

1508: Juan Ponce de Leon learned from the natives of Hispaniola that the island of Puerto Rico had many riches. For his help in Hispaniola, the rulers of Spain rewarded him with permission to explore Puerto Rico. Ponce de Leon established another Spanish colony there.

1509: Ponce de Leon was appointed governor of Puerto Rico. Natives there told him of an island called Bimini that contained the Fountain of Youth. According to a legend, anyone who drank from this fountain would never grow old.

1511: Because he was extremely brutal to Native Americans, Ponce de Leon was removed as governor.

1512: King Ferdinand II of Spain gave Juan Ponce de Leon permission to find and colonize Bimini (in the Bahamas).

1513: In March, Ponce de Leon left to find Bimini. He sailed with three ships: the *Santa Maria,* the *Santiago* and the *San Cristobal* and about 200 men. On March 27, Easter Sunday, his fleet sighted the eastern coast of Florida. On April 2, his ship landed on the east coast of Florida. He thought he had landed on an island. He named the land *Florida*, which means "flowery" in Spanish. He explored the Florida Keys and tried to prove that Florida was an island by sailing up the west coast. Ponce de Leon returned to Puerto Rico in September of 1513.

1514: Juan Ponce de Leon returned to Spain. The king gave him permission to colonize Bimini and Florida. The king gave Ponce de Leon the title Captain-General.

1515: Ponce de Leon returned to Puerto Rico and worked for the next six years to bring the natives under the control of the Spanish colony.

1521: Ponce de Leon took 200 men for a second expedition to try again to find Bimini and to colonize Florida. He landed on the west coast of Florida. While constructing houses, Ponce de Leon and his men were fiercely attacked by Native Americans. Ponce de Leon was wounded by an arrow and taken back to Cuba, where he died in July of 1521. Juan Ponce de Leon is buried in San Juan, Puerto Rico.

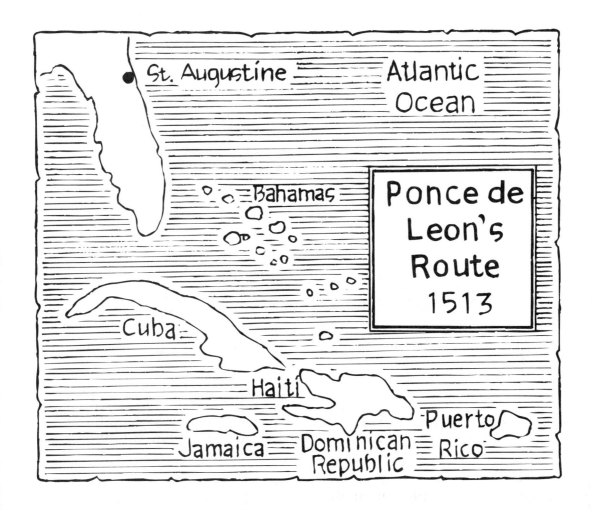

Vasco Núñez de Balboa

Spanish explorer in Central America

1475-1519*

*Sources differ on exact dates in the life of this explorer.

1475: Vasco Núñez de Balboa was born in Jerez de los Caballeros, Spain. He was the son of a poor nobleman.

1500*: Balboa sailed to Venezuela with an expedition led by Rodrigo de Bastidas. After exploring the southwestern portion of the Caribbean with Bastidas, he became a planter on the island of Hispaniola. He settled in Santo Domingo, the city first settled by Columbus in 1494.

1510: Balboa's plantation went bankrupt. He ran away to escape his creditors. He traveled as a stowaway on a ship headed for the new colony of San Sebastian on the coast of Colombia. It had been attacked by Native Americans and was left in rubble. Its founder had fled and abandoned the survivors of the colony.

Balboa persuaded the people to move to the Isthmus of Panama. (An *isthmus* is a narrow strip of land connecting two larger land masses.) Balboa was elected governor of the newly established settlement called Darien. He explored the inland areas and established Spanish rule. Unlike later conquerors, Balboa used negotiation, rather than force, in dealing with the Native Americans.

Balboa learned of a great ocean beyond the mountains and asked Spain to send more people to help him explore the area. Spain organized an expedition, but Balboa was not given the command. The king sent Pedrearias Davila to be commander and governor instead.

1513: Balboa was falsely accused of treason by his enemies in Spain. They turned the king against him. Balboa wanted to be on good terms with the king, so he searched for the great sea on the other side of the isthmus himself. He took 190 Spaniards and 1000 natives with him. He went through some of the thickest jungles on the continent. It took his group 24 days to go 45 miles through the dense jungle.

On September 29, Balboa and his group reached the sea. Because the men were facing south as they first viewed the ocean, they named it the South Sea. (It was renamed the Pacific Ocean by Magellan in 1520.)

Balboa sent news to the king of Spain of his discovery. He also sent gifts of gold and pearls he had found.

Balboa returned to Darien where he and Davila were in conflict immediately. Balboa was given permission to explore the South Sea, but then he was summoned home. He was arrested on the false charge of instigating a rebellion.

1519: Balboa was found guilty and beheaded.

Today, Panama honors Balboa by naming its monetary unit after him.

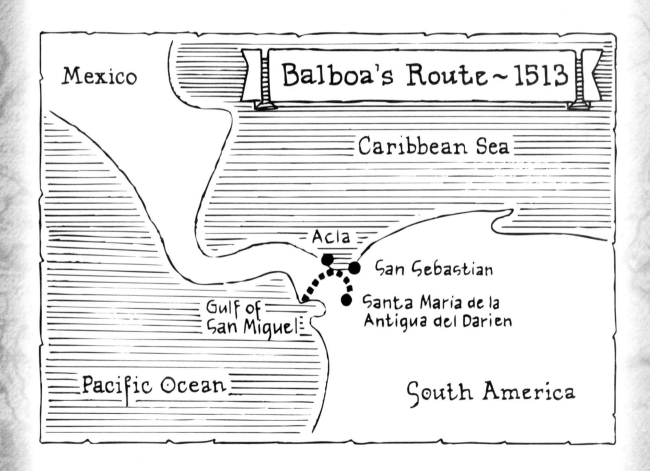

Francisco Pizarro

Spanish conqueror, explorer and governor

1475*-1541

*Sources differ on exact dates in the life of this explorer.

1475: Pizarro was born in Trujillo, a small town near Caceres, Spain. He was raised by his grandparents. They lived in poverty, and he herded swine as a youth. It appears Pizarro never learned to read or write.

1502: Pizarro traveled to the Caribbean and lived on the island of Hispaniola with a relative.

1510: He was part of an expedition to Colombia.

1513: Pizarro traveled with Vasco Núñez de Balboa on an expedition that ended in the discovery of the Pacific Ocean. He was second in command under Balboa.

1519-1523: Pizarro served as mayor of the town of Panama.

1523: Pizarro learned of a large and wealthy Indian empire to the south. We know it as Peru. He requested the help of two friends to go with him to explore and conquer the land. One of these friends, Diego de Almagro, provided equipment. The other friend, Hernando de Luque, was the vicar of Panama. He supplied the necessary money for the trip.

1525: On their first excursion, Pizarro's team traveled as far as the San Juan River in Colombia before turning back. The first expedition ended in disaster after two years of suffering and hardship.

1526-1528: Pizarro tried a second expedition to the Indian empire. It was going a little better, so he sent Almagro back to Panama for more help. Instead of sending help, the governor of Panama sent ships to bring back Pizarro and halt his expedition. Pizarro refused to return. Pizarro drew a line in the sand and asked all who wanted to share in his venture to join him. Thirteen men crossed the line and joined him. Pizarro's friends persuaded the governor to lend them one vessel. Pizarro used it to explore the coast of Peru. He collected information about the empire of the Incas. Then he sailed back to Spain to ask for the authority to conquer Peru.

1529: Pizarro was given permission from the Spanish government to conquer Peru. He was made captain-general. Almagro was made marshal.

1530: Pizarro left Spain on January 19 for Panama.

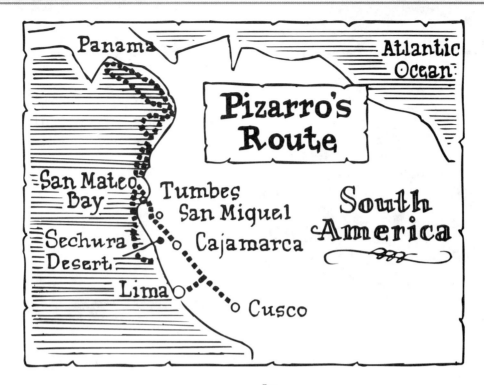

1531: Pizarro sailed from Panama with three vessels, 183 men and 37 horses. Almagro was to follow with reinforcements. Finally, after seven years of disappointment and hardship, Pizarro began the conquest of Peru. He spent a year conquering the settlements along the coast of Peru.

1532: Pizarro and his Spaniards began the march inland. In November, they entered the city of Cajamarca. Pizarro met with representatives of Atahualpa, the Inca emperor. The emperor accepted an invitation to visit Pizarro. Many unarmed Incas went with Atahualpa. Pizarro's men were armed and waiting. Pizarro asked Atahualpa to convert to Christianity or to accept the Spanish king as his ruler. Atahualpa refused. Pizarro and his men took the Inca emperor prisoner. They killed 2000 Incas.

As a ransom for his release, Atahualpa offered Pizarro a room full of gold as high as a man could reach. He would give them twice as much silver as well. Pizarro accepted the ransom and Atahualpa was released.

1533: Pizarro went back on his word and had Atahualpa executed. Pizarro then marched up to Cuzco, the Inca capital, and made Manco, Atahuallpa's brother, a ruler in name only. Pizarro was named a marquis by the Emperor Charles V.

1535: Pizarro founded the city now known as Lima. This was the seat of his new government.

1537-1538: Pizarro and Almagro quarreled about the territory each of them was to govern. The power struggle led to the War of Las Salinas in 1538. Pizarro's supporters captured and executed Almagro.

1541: On June 26, Almagro's followers assassinated Pizarro.

In 1984, it was confirmed that the remains of Pizarro were in a crypt in the Cathedral of Lima, Peru.

Ferdinand Magellan

Portuguese explorer who sailed around the world

1480*-1521

*Sources differ on exact dates in the life of this explorer.

1480: Fernao de Magalhaes was born in Oporto, Portugal. (*Ferdinand Magellan* is the English spelling of his name.) Ferdinand was the son of a Portuguese nobleman. He became a page at the Portuguese court.

1505-1510: Ferdinand sailed with a fleet carrying the first Portuguese viceroy to India. He then served with the fleet in the exploration and conquest of the East Indies. He was wounded twice in battle. He took part in expeditions that captured the kingdom of Malacca in the Malay Peninsula of southeastern Asia. He explored Indonesia as far east as the Spice Islands (Moluccas). By 1510, Magellan was promoted to the rank of captain.

1512: Magellan returned to Portugal.

1513: Magellan fought the Moors in Morocco. He was wounded again and was left with a permanent limp. King Manuel of Portugal became unhappy with Magellan (probably because of financial dealings) and canceled a promotion. Later the king denied Magellan's request for a fleet to sail west. Magellan wanted to prove that the Moluccas could be reached by sailing west.

1517: Magellan renounced his Portuguese citizenship and went to Spain to seek support for his plan from King Charles I.

Magellan believed there was a passage to the west through or around South America. He felt that such a route would be of great value to the Spanish, who wanted to share in the profits from the Spice Islands. The Portuguese claimed that all the islands of the Far East lay in the portion of the globe assigned to them by Pope Alexander VI, but Magellan claimed that many of them, including the Spice Islands, actually were in Spain's territory.

Magellan's Route—1519-1522

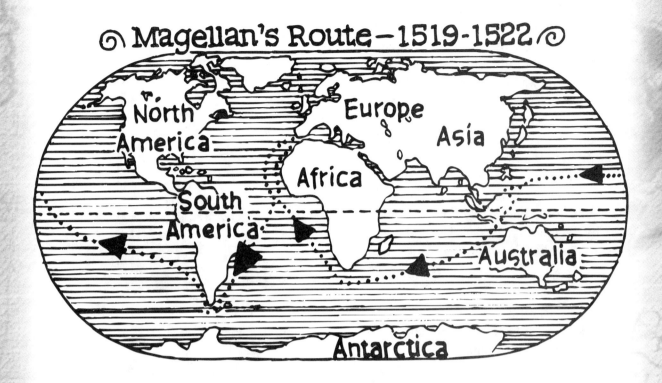

1519: The king of Spain finally agreed to Magellan's proposal. Magellan sailed from Spain with five vessels and about 250 men for his westward exploration. He sailed across the Atlantic Ocean and down the coast of South America until the southern hemisphere winter forced him to take shelter in April of 1520. He lost one ship when it was driven ashore while on an exploration. During the layover, Magellan's Spanish officers attempted a mutiny, but he put the mutiny down by force.

1520: Magellan resumed his voyage in August or September. On October 21, he sighted what he guessed was the strait around the southern tip of South America. He sent two ships ahead. They reported the strait led to an ocean. The rest of Magellan's fleet proceeded. The "ocean" turned out to be only a large bay in the strait.

Magellan continued to battle his way through the newly discovered strait. One vessel was wrecked, and another one returned to Spain. For more than a month he battled his way through the stormy 360-mile water passage. On November 28 he reached the ocean that Balboa had discovered seven years before. The route he used is now called the Strait of Magellan.

1521: Because it looked so calm, Magellan named Balboa's ocean the Pacific. The voyage across the Pacific started out well, but after a month of sailing the food ran low, and drinking water was scarce. Many of the crew died of scurvy. Finally, after about 100 days, the fleet arrived at the Mariana Islands in the western Pacific. Next, they traveled to Mindanao in the Philippines, where Magellan converted two local rulers to Christianity. He went on to Cebu Island where he made more converts. Magellan sailed from Cebu to the nearby island of Mactan. He and his crew became involved in a fight with some natives. Magellan was killed there on April 27, 1521.

1522: Juan Sebastian del Cano, one of Magellan's sailors, led the remainder of the voyage. One ship was abandoned because not enough sailors were left to handle three vessels. He brought the fleet to its goal, the Spice Islands. He took on a large cargo of cloves. Another ship started to leak and had to be left behind.

Only one ship, the *Victoria,* completed the three-year journey around the globe. Its cargo of cloves from the Spice Islands sold for such a high price that even though four of five ships were lost, the voyage earned a profit.

Magellan succeeded in sailing around the world before his death, but not in a single voyage. Ferdinand Magellan's greatest achievement was to confirm that the Earth is round. He showed that the world's oceans were connected, and he measured the Earth's circumference. His voyage laid the foundation for trade across the Pacific.

Hernando Cortez or Hernán Cortés

Spanish explorer and conqueror in Mexico

1485-1547

1485: Hernando Cortez was born in Medellín, Extremadura, Spain.

1499: Cortez was sent to Salamanca University at age 14 where he studied law.

1501: He quit law school two years later and then wandered for a few years.

1504: Cortez decided to sail for the Americas. He sailed to Santo Domingo (now the capital of Dominican Republic).

1511: Cortez joined the Spanish soldier Diego Velázquez, and took part in the conquest of Cuba. He then became the mayor of Santiago de Cuba.

1518: Cortez persuaded Válesquez, who had become governor of Cuba, to give him command of an expedition to Mexico.

1519: Cortez set sail from Cuba with 600 men. Válesquez cancelled the commission after becoming suspicious that Cortez would not submit to his authority. Cortez sailed anyway. He went along the Yucatan coast of Mexico and took over the town of Tabasco.

He took numerous captives, including one named Malinche, who became his mistress. She served as his guide and interpreter. The group of Spaniards then found a better harbor and established the new city of Veracruz. Cortez learned from the people of Tabasco of the Aztec empire led by Montezuma II.

Some crew members were opposed to Cortez rejecting the authority of Válesquez, so he destroyed his fleet in order to keep his men from deserting him. Cortez marched inland and formed an alliance with the Indians of Tlaxcala, who were the enemies of the Aztecs.

Cortez posed as Quetzalcoatl, who was a legendary god-king. This confused Montezuma. Cortez marched against the Aztecs and entered the capital city, Tenochtítlan, unopposed. He seized Montezuma as a hostage and made his own headquarters in Tenochtítlan.

Leaving explorer Pedreo de Alvarado in charge of his headquarters and 200 troops, Cortez traveled to the coast to defeat a Spanish rival, Narvaez. When he returned to Tenochtítlan, Cortez found the city in revolt. Alvarado had imposed harsh rules on the Aztecs. The Aztecs were revolting against the Spaniards, and they wanted the return of their leader, Montezuma.

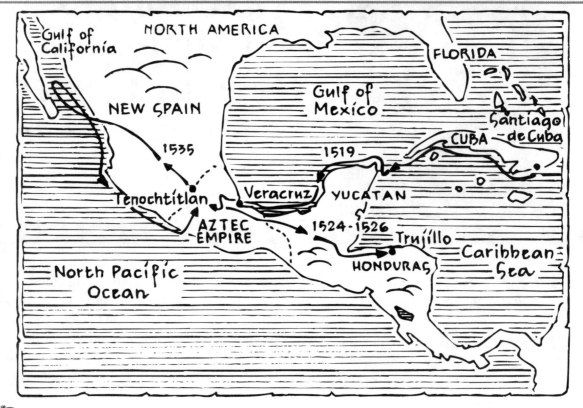

1520: The Spanish suffered heavy losses on the night of *Noche Triste* ("sad night") of June 30. They were forced to withdraw from Tenochtítlan.

1521: Cortez recaptured Tenochtítlan, but only because an epidemic had killed many of the city's defenders. Cortez built Mexico City on the ruins of Tenochtítlan. It became the primary European city in America. Cortez returned to Spain several times.

1523: Cortez was named the governor and captain general of New Spain.

1524-1526: Cortez took an expedition to Honduras.

1528: Cortez was ordered to return to Spain and to renounce his title in Mexico because the Spanish officials were fearful of his intent. He appealed to the king and was made marquis of the Valley of Oaxaca in southern Mexico. He was not restored as governor, however. He married the daughter of the Count of Aguilar.

1530: Cortez returned to Mexico. His activities were monitored continually.

1536: He discovered Baja, California, in northwest Mexico, and he explored the Pacific coast of Mexico.

1539: The Spanish explorer, Coronado, was granted rights to seek the Seven Cities of Cibola. Cortez went back to Spain to complain. He was received with honor but did not recover his property.

1541: Cortez participated in the unsuccessful attack on Algiers. He lost a large part of his fortune and was shipwrecked. He retired on a small estate near Seville.

1547: Hernando Cortez died on his estate in Seville, Spain.

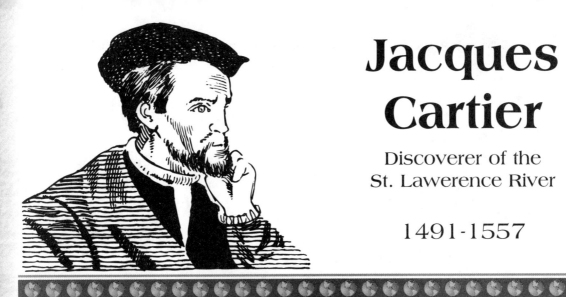

Jacques Cartier

Discoverer of the St. Lawerence River

1491-1557

1491: Jacques Cartier was born in Saint-Malo, France, on December 31.

1534: King Francis I chose Cartier to lead an expedition to find the Northwest Passage to China. In April, Cartier left Saint-Malo with two ships. He reached northern Newfoundland and Labrador in May. He passed through the Strait of Belle Island and explored the Gulf of St. Lawrence. He did not reach China.

1535-1536: Cartier sailed on his second voyage with three ships. He sailed up the St. Lawrence River as far as Stadacona, which is modern Quebec. He then went to Hochelaga and climbed the hill to see the Ottawa River. Cartier called the hill Mont Real (Mount Royal), which today is called Montreal. Cartier continued up the river until he was stopped by the Lachine Rapids. He spent the winter at Stadacona, where 25 of his 110 men died of scurvy. The discovery of a medicinal brew of white cedar saved the rest of his men. Cartier returned to France in July 1536.

1541-1542: On his third journey, Cartier took colonists in five ships. He was required to sail under the command of de Roberval. They traveled to Cape Rouge, near Quebec. Cartier founded a settlement there. He also found quartz, which he mistook for diamonds. He found iron pyrite, which he thought was gold. He spent the winter there. At least 35 members of his crew were killed by Iroquois.

1542: Cartier returned to France. The colony, under de Roberval, did not succeed. For a time, France lost interest in Canada. Cartier retired, but served as an adviser on navigation.

1557: Cartier died in Saint-Malo, France, on September 1.

Cartier did not find a sea passage to the East Indies through North America. Instead, he found the St. Lawrence River. He opened Canada to European settlement. His explorations were the basis for France's claim to Canada.

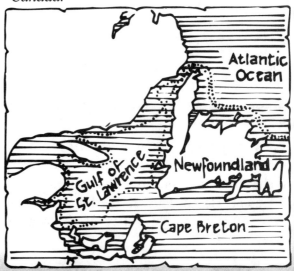

Hernando de Soto

Spanish explorer of Florida

1500-1542*

*Sources differ on exact dates in the life of this explorer.

1500: Hernando de Soto was born in Barcarroto, Spain, about 1500.

1514: De Soto went to Darien (now Panama) as an aide to the governor, Pedrarius Davila.

1519-1530: De Soto took part in several expeditions to Central America.

1532: De Soto joined with Pizarro in the conquest of the Inca Empire in Peru. The Peruvian treasures made him very rich. He was the first European to meet the ruler Inca Atahuallpa. (Pizarro had Atahuallpa imprisoned and then executed. De Soto strongly disapproved of Pizarro's cruelty.)

1536: De Soto returned to Spain where he married Davila's daughter, Isabel.

1537: Holy Roman Emperor Charles V (who was also king of Spain) named de Soto governor of Cuba and Florida. He was allowed to explore and conquer for Spain.

1538: De Soto left Spain and sailed to Havana with a force of 1000 men.

1539: De Soto sailed from Havana, Cuba, for Florida with about 600 men. They landed near Tampa Bay and marched along the Gulf of Mexico toward Georgia. When his men were discouraged or supplies ran low, de Soto told his men of the riches ahead. They met many Indian tribes as they traveled. De Soto forced them to furnish supplies and tell them where gold was hidden. De Soto's tactics caused many needless battles. About 70 Spaniards were killed and de Soto was severely injured.

Over four years, de Soto and his men explored about 350,000 square miles of the southeastern United States. They went as far west as Texas and as far north as the northern boundary of Arkansas.

Holy Roman Emperor Charles V

1541: De Soto's team became the first Europeans to see the Mississippi River, probably near present-day Memphis, Tennessee. De Soto's men built boats and crossed the Mississippi River. Everywhere de Soto searched, the native people told him that gold was "just ahead."

1541-1542: De Soto's army spent the winter near the junction of the Canadian and Arkansas rivers in Oklahoma.

1542: De Soto led his men, with dampened spirits, southward to the Mississippi River, where he died of a fever. De Soto had told the local people that he was immortal. So when he died, his aide, Moscoso, sank his body in the river to keep the local people from learning of his death.

De Soto and his men never found the gold they had searched for in the southeastern United States. They left behind many dead friends and enemies. They also imported European diseases to the western hemisphere that killed many natives who lacked resistance to the diseases.

In 1948, the de Soto National Memorial was established near Saint Petersburg, Florida, commemorating his landing in Florida.

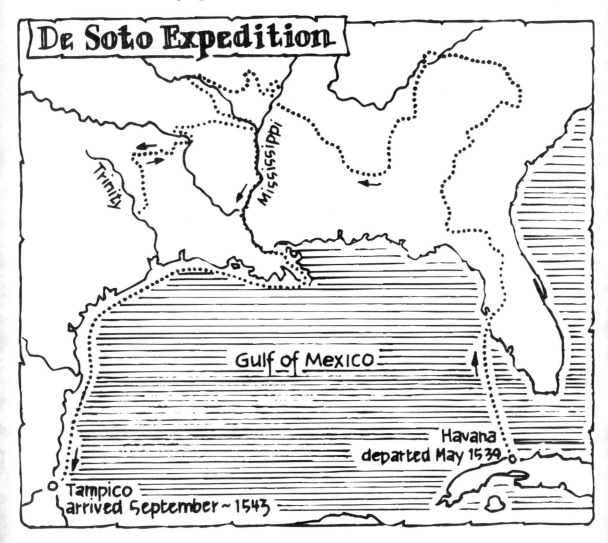

Francisco Coronado

Spanish conqueror and explorer of North America's Southwest

1510-1554*

*Sources differ on exact dates in the life of this explorer.

1510*: Francisco Vazquez de Coronado was born into a noble family in Salamanca, Spain. As a young man, he became friends with Antonio de Mendoza, one of the king's favorites.

1535: Coronado's friend, Mendoza, was appointed viceroy of New Spain (Mexico). Coronado went with him to America. While in Mexico City, Coronado married the wealthy Beatriz Estrada.

1538: Mendoza made Coronado governor of New Galicia, a province in western Mexico. There Coronado heard the tales of the Spanish explorer, Cabeza de Vaca, about the Seven Golden Cities of Cibola. These cities were believed to be extremely rich Native American settlements and were located northeast of the Mexican province where Coronado lived. Mendoza chose Coronado to lead an overland expedition to explore the seven cities, seize the treasure and conquer the region for Spain.

1540: Coronado led his party from New Galicia in search of the Seven Golden Cities of Cibola. His army consisted of approximately 300 Spaniards, Native Americans and slaves. Joining them were herds of cattle, sheep and pigs. A fleet in search of an inland waterway to Cibola took a parallel course along the coast.

In July, Coronado's group arrived at the first of the promised cities. (The cities were actually the Native American pueblos of present-day Zuni in western New Mexico.) From here, Coronado sent out scouting parties. The men in one party, led by Garcia Lopez de Cardenas, became the first Europeans to see the Grand Canyon. (Because of the inaccessibility of the canyon, it was another 300 years before the canyon was explored.)

Another of Coronado's scouting parties found more pueblos in a fertile area of the Rio Grande valley. But still they found no gold or riches. The men in Coronado's expedition spent the winter there, near what is now Santa Fe, New Mexico.

1541: A passing slave told the men of a new land to the northeast whose capital city, Quivira, was very rich. This slave guided Coronado and 30 of his men eastward toward this city. They crossed the Rio Grande and the Great Plains of northern Texas, where they saw American bison and described them for the first time. Finally, after months of travel, they found Quivira in what is now central Kansas. They found only a village of the Wichita tribe. The slave confessed he had made up the whole story about its riches. He was executed. Coronado then returned to the Rio Grande valley. He and his men spent a second winter there before returning to Mexico City.

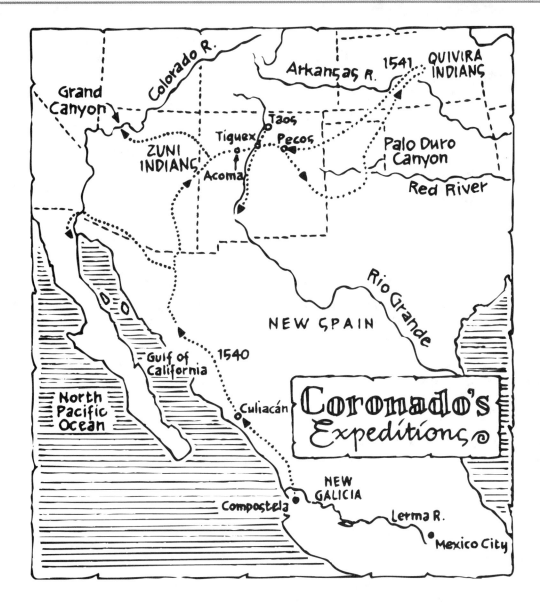

1542: Coronado led about 100 men into Mexico City. The remainder of the survivors straggled in over the next months.

1544: An official inquiry was made into Coronado's journey. He was charged with corruption and negligence. He was removed as governor of New Galacia. Coronado returned to a quiet life in Mexico City.

1554: Francisco Coronado died on September 22.

The account of Coronado's explorations is valued for its unique description of the southwestern United States. It was published in a U.S. government report in 1896. The Coronado National Memorial honoring Coronado's expedition was established in 1952 near Bisbee, Arizona.

Sir Francis Drake

English navigator and pirate

1540*-1596

*Sources differ on exact dates in the life of this explorer.

1540: Francis Drake was born near Tavistock, in Devonshire, England. His family was a poor, farming family of devoted Protestants. As a boy, Drake worked as a sailor. From the age of 13, Drake worked in trade along coastal areas.

1565: By this date, Drake was traveling to Guinea, on the west coast of Africa, and to the Spanish mainland to obtain slaves.

1566: He sailed to the West Indies on a slave-trading expedition. (The West Indies refers to the region around the Caribbean Sea.)

1567: Drake was given his first command of a ship. His 50-ton vessel, the *Judith*, was part of the fleet led by his cousin, Sir John Hawkins. They went on a slave-trading voyage in the Gulf of Mexico. The fleet was attacked by the Spanish at San Juan de Ulua (Vera Cruz, Mexico). All but two of the ships were lost, and Drake lost nearly everything he owned. The ships led by Drake and Hawkins were the only ones to escape and return to England. It is said that Drake never forgave the Spanish for this, or for their cruel treatment of their English prisoners. He became determined to fight against Spain.

1570: Drake explored the West Indies. He looted Spanish settlements and demolished Spanish ships.

1571: Drake explored the West Indies again and took large quantities of silver and gold.

1572: He commanded two ships in attacks against the Spanish ports in the Caribbean Sea. He made a daring march across the Isthmus of Panama. From a high tree, Drake saw the Pacific Ocean for the first time.

1573: Drake returned to England with a cargo of Spanish silver. Because of his triumphs and valuable cargo, Drake became a popular hero.

1573-1576: Drake was sent to Ireland to help subdue the Irish rebellion.

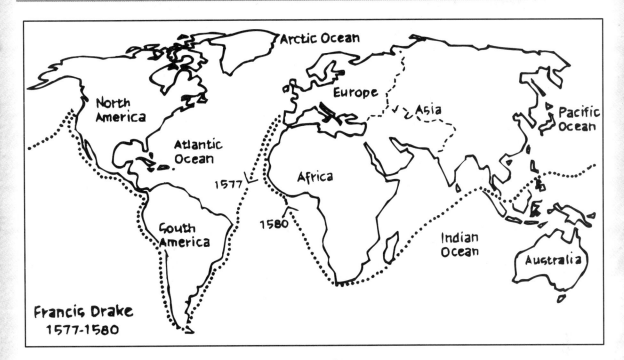

1577: In order to end the Spanish control of the trade in the Pacific Ocean, Queen Elizabeth I secretly appointed Drake to make a voyage around the world. This had been done only once before, by Ferdinand Magellan. Drake launched from Plymouth, England, with five ships and 166 men. It proved to be a difficult trip. At the southern tip of South America, Drake had to overcome a mutiny. He abandoned two small store ships.

1578: The three remaining ships went through the Strait of Magellan. After 16 days, they sailed into the Pacific Ocean and headed north. Severe storms lasting 50 days destroyed one ship. Another ship abandoned the voyage and sailed back to England. Drake completed the voyage with just one ship, the *Golden Hind*. Drake attacked the Spanish colonies in Chile and Peru. He continued north, possibly as far as the present U.S.-Canadian border. He was looking for an eastward passage back to the Atlantic Ocean. When he did not find one, he sailed south and landed at a bay that is now called Drake's Bay. It is just north of San Francisco. Drake claimed the land for England.

1579: Drake left the coast of California and sailed west across the Pacific Ocean. He did not see land for 68 days. Drake stopped at some islands in Indonesia, and then continued around the southern tip of Africa.

1580: After nearly three years, Francis Drake reached Plymouth, England, in September. He carried silver and Spanish treasure. Recognized as the first Englishman to go around the world, Drake was knighted seven months later by Queen Elizabeth I for his accomplishments.

1581: Drake became mayor of Plymouth.

1584-1585: Drake served as a member of Parliament.

1585: He sailed with a large fleet for the West Indies. He raided many Spanish settlements, including one in present-day Florida. Before returning to England, Drake sailed north to Roanoke Island, off the coast of what is now North Carolina. He retrieved nearly 200 disheartened colonists and brought them back to England.

1587: Drake led a raid on the Spanish port of Cadiz. He destroyed many stores and Spanish ships. Drake destroyed so many Spanish ships that the attack by the Spanish Armada on England had to be delayed for a year.

1588: Drake served as vice admiral of the English fleet. He helped lead the fleet in defeat of the Spanish Armada.

1589: Drake attempted to defeat the remaining Spanish ships, but was unsuccessful. He returned to Plymouth and to Parliament.

1595: The queen sent Drake and Hawkins to the West Indies to battle the Spanish again, but the mission was a failure.

1596: Drake and Hawkins both became ill with dysentery, an infection of the intestines. They died in the Caribbean and were buried at sea off the coast of Panama.

More than any other shipman, Drake helped England become a powerful force on the seas.

Samuel de Champlain

Explorer, mapmaker and founder of Quebec

1567-1635*

*Sources differ on exact dates in the life of this explorer.

1567: Samuel de Champlain was born in Brouage, France. His father was a naval captain, which gave Samuel opportunities to train as a navigator, geographer and mapmaker.

1595: Champlain's career began in the military.

1598: By the end of the Wars of Religion in France, Champlain was the captain of a company stationed at Blavet, France.

Unemployed after the war, Champlain left for Spain. He sailed on a ship belonging to his uncle that had been commissioned by the king of France. After arriving in Cádiz, Spain, Champlain was granted permission to leave for the West Indies. He spent two years in the Spanish colonies in Peru and Mexico. He gained a reputation as a fine navigator during this voyage. King Henry IV was impressed by his reports from Central America. He made Champlain the royal geographer.

1603: King Henry wanted to establish French settlements in Canada. Champlain joined a group that was sent to find sites in Canada for settlement and trading. He traveled as a royal geographer on an expedition with fur traders. This was his first of 12 voyages to North America. The group traveled to Tadoussac, at the mouth of the Saguenay River. It was a trading center for the French and Indians. Champlain traveled along the Saguenay and gathered information on the northeastern part of the continent. From his travels, he drew a map that was quite accurate. His map included a large bay to the north (Hudson Bay). It showed water to the west, which he found out later was the Great Lakes. He believed the lakes were connected to the Pacific Ocean.

1604: Champlain made his second trip to North America. This time he took with him a group of colonists. They built homes and a storehouse at the mouth of the St. Croix River in New Brunswick. The settlers lived through a hard winter. In the spring they moved to a better location in Nova Scotia. During this time, Champlain explored the area to the south going as far as Cape Cod.

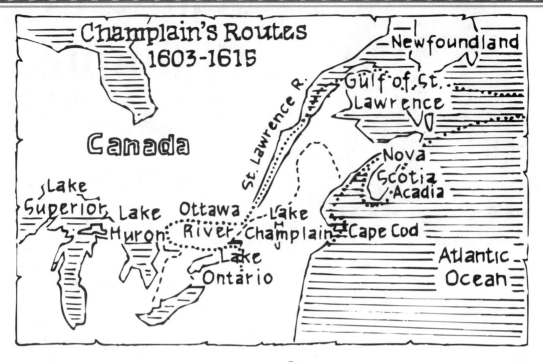

1608: While in Canada, Champlain was given permission to lead another expedition. He founded Quebec and made friends with the Huron tribes. Quebec became the first permanent white settlement in the region that is now Canada. It is the oldest city in the western hemisphere north of Saint Augustine, Florida.

1609: Champlain joined the Hurons to fight the Iroquios in New York. He discovered Lake Champlain. Near that lake, he defeated the enemy with gunfire. Because of that, the Iroquois were bitter enemies of the French, on and off, for 90 years.

1612: Champlain was given the title of Lieutenant of the Viceroy of New France. At this point, Champlain determined "to explore and map the continent, to find a water route to the Pacific, and to convert the indigenous peoples to Christianity." To fund this venture, he made alliances with northern and western nations to trade furs.

1613: He explored the Ottawa River. This would become the main route of travel to the west.

1615: Champlain reached the Georgian Bay and Lake Huron.

1616: After this date, Champlain acted as a leader in Quebec for various sponsors. He traveled to France often, seeking funds.

1629: The British seized Quebec, not knowing that a peace treaty with France had already been signed. The British took Champlain to England as a prisoner.

1632: Champlain was released by the British. The colony of Quebec was returned to the French.

1633: Champlain returned to Quebec as governor.

1635: Champlain died on December 25 in Quebec.

Bartholomew Gosnold

Explorer of Cape Cod and Jamestown

1572-1607

1572: Bartholomew Gosnold was born in Suffolk, England. Gosnold later graduated from Cambridge University and studied law at Middle Temple.

1600: Queen Elizabeth I chartered the East India Company so that England could gain a foothold in the Far East.

1602: Gosnold commanded the ship *Concord* and pioneered a direct route from the Azores to New England. Gosnold was the first Englishman to explore the east coast of the current United States from Maine to Massachusetts.

He named Martha's Vineyard after his daughter. He named Cape Cod for the fish he found there. Gosnold established a small post on Elizabeth Island, which he named after the queen. (Today, the island is known as Cuttyhunk Island.) The post, now part of the town of Gosnold, was abandoned when the crew voted to return to England instead of spending the winter there.

Gosnold also explored Nantucket Sound and Narragansett Bay. He returned to England with furs, lumber and sassafras that he had received in trade from Native Americans. Also upon his return to England, Gosnold actively encouraged others to begin colonies in the areas he had explored.

1606: Gosnold helped gain exclusive rights for a company to settle Virginia.

1606-1607: Gosnold was chosen to command the *God Speed,* one of three ships that took English settlers to Jamestown, Virginia. To travel with him, Gosnold recruited John Smith, Edward Wingfield, a brother, a cousin and some members of his 1602 voyage.

1606: Jamestown, Virginia, was founded by John Smith, Gosnold and others. Gosnold was popular among the colonists. He helped design the fort that held the first colony, and he was appointed to its council. Just four months after the ships landed at Jamestown, Bartholomew Gosnold died of illness.

In 2005, archaeologists announced that they believed they had found Gosnold's grave. They began to try to verify that by using DNA.

Henry Hudson

English navigator

1575-1611*

*Sources differ on exact dates in the life of this explorer.

1575*: Henry Hudson was born in England around 1575. The year of his birth is uncertain. Little is known of his life before 1607.

Because of the thriving trade in spices and silk between Europe and the Far East, many explorers searched for a Northeast or Northwest Passage. This passage would be a shortcut northward from the east coast of Europe. The desired route could then proceed (eastward) over the top of Europe and Asia. Or it could go (westward) over the top of North America. Either passage, if found, would provide a shorter, quicker way to the Pacific. Henry Hudson tried to find both passages.

1607: The English Muscovy Company hired Hudson to find a northeast passage. He commanded just one ship, the *Hopewell*. Hudson sailed from England to Greenland. He tried to find a passage to the Far East by way of the Arctic Ocean. He was blocked by ice.

1608: Hudson sailed in the same ship to search again for a passage. He sailed to Novaya Zemlya, islands north of Russia in the Barents Sea. He was again turned back by ice.

1609: The Muscovy Company withdrew its support of Hudson's voyages. So Hudson turned to a new company, the Dutch East India Company, for a ship and financial support. Hudson made his third voyage in his new ship, the *Half Moon*, as an employee of the Dutch company. With a mixed Dutch and English crew of about 18-20 men, he again sailed in search of a passage off Novaya Zemlya.

The weather was extremely cold. Hudson's crew mutinied, and Hudson headed west and south past Nova Scotia and down the North American coast. He believed that the Atlantic and Pacific Oceans were separated by a narrow strip of land, or isthmus. He explored down toward Virginia, looking for a passage across the continent. He then turned northward and sailed into a river that was later named the Hudson River, in his honor. Hudson sailed 150 miles, as far as Albany, New York.

Before the end of 1609, Hudson then returned to England where the men and the ship were seized by the government. Hudson was told he could only sail for the government of his birth.

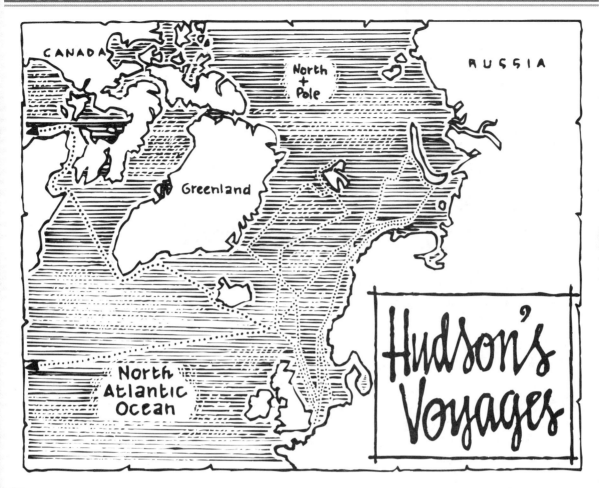

🧭 **1610:** Hudson took his fourth and final voyage under the sponsorship of a newly formed English company. His new ship was the *Discovery*. He took a crew of 20, along with his 12-year-old son. This time Hudson decided to search for the northwest passage. He reached the Hudson Strait by mid-year. Then he sailed into what is now Hudson Bay. He spent three months exploring the eastern islands and shores. By November his ship was frozen in. A harsh winter led to disagreements among the crew.

🧭 **1611:** By June of 1611, part of Hudson's crew mutinied. The crew put Hudson, his son and seven others in a small boat. The stranded explorers were left to freeze or starve. They were never seen again. Eight survivors from the crew reached England. They were tried for mutiny but not found guilty.

Although Henry Hudson did not succeed in finding a northeast or northwest passage to the Pacific, his four voyages greatly increased the knowledge at that time of North America and the Arctic.

Peter Minuit

Colonizer for the Dutch in North America

1580-1638

1580: Peter Minuit was born in Wesel, Germany, in about 1580.

1620s: Sometime in the early 1620s, Minuit moved to the Netherlands.

1626: Minuit joined the Dutch West India Company. He sailed to the company's settlement of New Netherland along the Hudson River in 1626. Minuit went to Manhattan Island and traded a few trinkets, valued at 60 Dutch guilders (about $24), with the Algonquin-speaking tribes. The tribes thought they were agreeing to share the land with the Dutch, but the Dutch considered it a purchase.

Minuit built a fort on the southern tip of the island called New Amsterdam. (Our modern New York City grew out of that establishment.)

1631: Minuit had differences with the company and was brought back to Holland. He later entered Swedish service.

1637: Minuit set out to establish a Swedish colony in America.

1638: Minuit built Fort Christina (named after the queen of Sweden) in what is now Wilmington, Delaware. The surrounding territory was called New Sweden. That same year, while traveling at sea to the West Indies, Peter Minuit was lost in a hurricane and died.

Abel Tasman

Dutch explorer of Tasmania and New Zealand

1603-1659*

*Sources differ on exact dates in the life of this explorer.

1603*: Abel Janszoon Tasman was born in approximately 1603 in the Netherlands.

1632: Tasman entered the service of the Dutch East India Company. He made his first expedition to the island of Seram in modern Indonesia. Later he sailed on several more voyages in the Far East.

1642: Tasman was chosen by the governor-general of his country, Anthony van Diemen, to lead an important expedition in the southern hemisphere. His assignment was to travel through the Indian Ocean and the South Pacific Ocean to find a passage by water to Chile. He was also to search for "Terra Australis," the legendary continent thought to exist in the southern hemisphere. (At that time a very large region of the hemisphere was still unexplored.)

Tasman left with two ships from Batavia (now Jakarta, Indonesia) on the island of Java, on August 14. He sailed first to Mauritius, then southward, and finally northward again, where he discovered a large island on November 24. He named it Van Diemen's Land, after the governor-general who had commissioned him for the expedition. But now the island is known as Tasmania, after the man who landed there.

Tasman continued eastward. In December, he arrived at New Zealand. He was still sure there was a passage to Chile, so he sailed northward once more. He explored Tonga and the Fiji Islands. He also explored New Guinea. On this voyage, Tasman circled Australia but never saw it. And although Tasman sailed to New Zealand in 1642, the people there remained undisturbed by Europeans until the arrival of James Cook in 1770.

1643: Tasman returned home on June 14.

1647: Tasman made one more exploration, taking a trading fleet to Thailand.

1648: Tasman commanded a war fleet against the Spanish in the Philippines.

1659: It is believed that Tasman died in October.

Louis Hennepin

Belgian explorer in the Mississippi River Valley

1626-1701*

*Sources differ on exact dates in the life of this explorer.

1626: Louis Hennepin was born in Ath, now in Belguim, on May 12, 1626.

As a young man, Hennepin joined the Franciscans, a part of the Roman Catholic church. He preached in Belguim, Luxembourg and the Netherlands.

1675: Hennepin sailed to Canada. He became a chaplain to Robert La Salle and a missionary to the Native Americans. (See also the biography of René-Robert Cavelier, sieur de La Salle.)

1676: Hennepin went to Fort Frontenac, presently the site of Kingston, Ontario. He began a mission among the Iroquois.

1679: Father Hennepin traveled with La Salle aboard the *Griffon* on the Great Lakes. They began near Niagara Falls and crossed Lakes Erie, Huron and Michigan. The expedition traveled across Illinois country and to the upper valley of the Mississippi River. They reached the Illinois River and built Fort Crevecoeur near present-day Peoria, Illinois. This was the first European settlement in Illinois.

1680: La Salle turned back to go for fresh supplies. Hennepin and his group were left to explore the upper Mississippi River. But they were captured by the Sioux. While traveling with the Native Americans, Hennepin became the first European to view the falls where Minneapolis, Minnesota, now stands. He named them the Falls of St. Anthony.

1681: Hennepin was rescued, and he returned to France.

1683: Hennepin became famous when the story of his journeys, *Description of Louisiana,* was published.

1687: After the death of La Salle, Hennepin published another book in which he claimed to have traveled down the Mississippi River before La Salle. This was not true, and it caused people to doubt his earlier book.

Hennepin decided not to return to America, which displeased the Franciscan authorities.

1701: Louis Hennepin died in Rome in about 1701. He was virtually unknown.

Hennepin's writings sparked interest in the interior regions of North America. They helped draw the French into the development of the Gulf Coast area. Hennepin County, which includes the city of Minneapolis, is named for Father Louis Hennepin.

Jacques Marquette

French explorer of the Mississippi River

1637-1675

1637: Marquette was born in Laon, France. He was the youngest of six children born to Nicolas and Rose Marquette. His family had been well-known in Laon since the 14th century. Jacques was often known as Père Marquette.

1654: Jacques was a thoughtful and gentle person, and at the age of 17 he decided to become a Jesuit priest. He went to Nancy, France, to study.

1656: Marquette began studying philosophy and taught at Jesuit schools in France. He hoped to someday become a missionary and travel overseas.

1666: His Jesuit superiors sent Jacques as a missionary to Quebec in New France (now Canada). He studied Native American languages for two years.

1668: Marquette founded a mission in Sault Sainte Marie (Michigan).

1669: He met the young Canadian explorer Louis Jolliet.

1669-1671: Father Marquette left the mission in Sault Sainte Marie in order to go to La Pointe mission in the Apostle Islands of Lake Superior. He spent 18 months there. He was visited by some Native Americans from the Illinois tribes. He became friends with them and wanted to start a mission among them. The Illinois people told Marquette of a great river they had crossed to reach him.

1671: A quarrel erupted with the Sioux. They forced Marquette and his converts to flee to Mackinac (Michigan). There he founded a mission at what is now St. Ignace, in Michigan's Upper Peninsula.

1672: Jolliet came to St. Ignace. He had been commissioned by the governor of New France to seek the great river of which the Indians spoke. He brought news that Father Marquette was to join him.

1673: In May, Marquette, Jolliet and five other men set off in two bark canoes. They paddled down Green Bay into the Fox River, then to the head of the Wisconsin River. On June 17, they reached the great river, the Mississippi River. Marquette became one of the first Europeans to see any part other than the lower Mississippi River. (Spanish explorer Hernando de Soto had explored the lower river about 130 years earlier.) Marquette preached to the Native Americans along the way.

Marquette's group traveled as far as the Arkansas River. They learned that the Mississippi eventually flowed into the Gulf of Mexico, not the Pacific Ocean. They knew if they kept going, they would reach Spanish territory. So Marquette, Jolliet and the others turned back north. They reached the Illinois River, then followed it into the Des Plaines River. Next they reached the Chicago River and took that to Lake Michigan.

Near the end of September, Marquette's group finally rested at the mission of St. Francis Xavier at De Pere (near modern Green Bay). They had traveled more than 2500 miles in less than five months. The journey took a toll on Marquette's health.

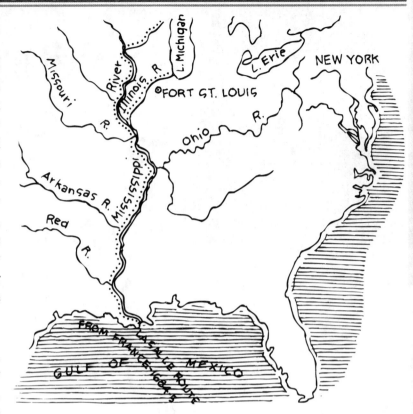

Marquette's journals of his exploration along the Mississippi River were first published in 1681, after his death. They include maps, descriptive notes and his correspondence. They give clear information about explorations in the Great Lakes and upper Mississippi regions. His account was of help to La Salle, who explored the Mississippi River to its mouth at the Gulf of Mexico in 1682.

1674: Marquette stayed at De Pere for more than a year. While he recovered physically, he wrote a journal of his voyage on the Mississippi. In October, he set out to start a new mission among the Illinois Native Americans.

1675: Marquette and two companions set up a camp at the mouth of the Chicago River and were forced to spend the winter there. He was visited by groups of Illinois Native Americans. By early spring, Marquette thought he was well enough to continue back to the village on the shore of the Illinois River. He preached to a large crowd on Easter Sunday. But his health worsened. Marquette tried to return to his mission at St. Ignace, but he collapsed and died on the way on May 18. He may have been buried in either Frankfort or Ludington, Michigan.

René-Robert Cavelier, sieur de La Salle

French explorer of the Mississippi River

1643-1687

1643: René-Robert Cavelier, sieur de La Salle, was born on November 22 in Rouen, France. He was the son of a rich merchant and was educated by Jesuits. He studied logic, physics and mathematics.

1666: La Salle immigrated to Canada. He was given land at Lachine, near Montreal, from the Seminary of St. Sulpice where his older brother was a priest. He was more interested in becoming a fur trader than he was in farming the land. Every spring, Native Americans came by canoe in large numbers to trade bales of furs for French cloth, brandy, guns and more. Then the Native Americans disappeared into the wilderness until the following year.

La Salle learned the language of the Iroquois, as well as other dialects. He heard from the Native Americans that there was a large river south of the Great Lakes that ran to "the Vermilion Sea." He thought this sea might be the Gulf of California. If it were, this great river would be an ideal route to China.

1669-1670: La Salle sold his land to pay for an expedition. He traveled down the St. Lawrence River to Lake Ontario and along the lake's southern shore. The records from the remainder of that exploration were lost. When he returned to Canada, Count Louis de Frontenac was in power.

1671: La Salle claimed that he had discovered the Ohio River.

1673: By this time, Louis Jolliet and Father Jacques Marquette had explored the Mississippi far enough to prove that it emptied into the Gulf of Mexico. Frontenac and La Salle immediately suggested that France build forts and trading posts along the Great Lakes and the Mississippi in order to keep the region and its fur trading business for France. The forts would protect the route from the Iroquois. They began by building Fort Frontenac near Kingston.

1674 and 1677: Frontenac sent La Salle to France as his representative. He explained to the French government why Fort Frontenac was needed. La Salle successfully made his case. He was given a monopoly on trade in the area and was given the title of *sieur* ("sir" in English). After his second trip to France, La Salle returned with Italian explorer Henri de Tonti, who became his business partner.

1678-1679: An advance group built a fort at the Niagara River and started to build a 40-ton ship called the *Griffon*.

1679: In August, La Salle, Tonti, Father Louis Hennepin and others left from near Niagara Falls. They sailed across the Great Lakes toward Green Bay. It was the first voyage ever made by a ship on the Great Lakes. They reached Green Bay in September and sent back the *Griffon*, filled with furs. In December, they built Fort Miami on Lake Peoria. (See also the biography of Louis Hennepin.)

1680: La Salle built Fort Crevecoceur. (In English, this name means "Fort Heartbreak.") From there, he sent Father Louis Hennepin and two companions to explore the upper Mississippi. La Salle left Tonti in charge of Fort Crevecoeur while he returned to Fort Frontenac. He learned the *Griffon* had never arrived there. When he traveled back westward, he discovered the Iroquois had destroyed Fort Crevecoeur. Tonti and his men had disappeared. La Salle traced Tonti to Mackinac.

1681: La Salle organized the Illinois tribes to resist the Iroquois.

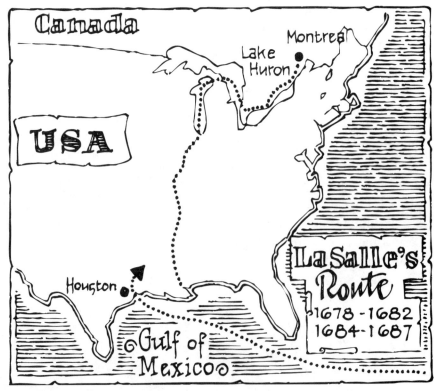

1682: La Salle's expedition followed the Illinois River and the Mississippi River to the Gulf of Mexico. They claimed all the land drained by the great river for the king of France, Louis XIV. La Salle named the region *Louisiana*. He began building forts in the new territory.

1683: La Salle returned to Canada and learned that Frontenac had been called back to France. His replacement had canceled La Salle's rights. La Salle went to France and persuaded King Louis XIV to restore his rights. The king helped him obtain four ships and about 400 people to establish a French colony at the mouth of the Mississippi River.

1684-1687: LaSalle's final expedition by sea was filled with failure. The naval commander chosen by the king, Beaujeu, constantly opposed LaSalle. In the West Indies La Salle became sick, he lost one ship to pirates, and many in his group deserted him. When he reached the Gulf of Mexico, LaSalle was unable to find the mouth of the Mississippi River. He landed at Matagorda Bay, Texas. In March of 1685, Beaujeu deserted LaSalle, and left him with just one small ship. LaSalle continued to look for the Mississippi River. But in 1686, LaSalle's ship was wrecked, and only 36 of his men remained.

1687: In January, La Salle took half of his men on an overland trip to reach Tonti in Illinois. In March, three of those men killed LaSalle.

*In 1995, one of La Salle's supply ships, the **Belle**, was discovered in Matagorda Bay. It had disappeared when it sank during a storm in 1686.*

Louis Jolliet

French-Canadian explorer of the Mississippi River

1645-1700

1645: Louis Jolliet was born on September 21 near the city of Quebec, Canada.

Jolliet was educated in a Jesuit seminary and studied to become a priest. Before taking his final vows, he changed his mind and went to France for a year to study science.

1668: Jolliet became a fur trader among the Native Americans. He traveled through wilderness areas around the Great Lakes. He became a skilled mapmaker, and he learned the languages of Native Americans.

1669: Jolliet met Jacques Marquette.

1672: The governor of New France chose Jolliet, who was already familiar with the region, to lead an expedition to find the great river in the west. At that time, it was only known by rumor. Marquette and five other men were chosen to join Jolliet.

1673: Jolliet's group left St. Ignace (Michigan) on May 17. They crossed Lake Michigan, went up the Fox River and down the Wisconsin River. On June 17, the men entered the Mississippi River. They followed the great river southward to just below the mouth of the Arkansas River before turning back. Marquette remained at Lake Michigan while Jolliet continued on to Quebec.

1674: On his journey to Quebec, Jolliet's canoe tipped over, and he lost his maps and papers. He was able to replace many of them from memory. Jolliet was granted the feudal rights to several islands in the lower St. Lawrence River.

1675: Jolliet was married. He made his home on Anticosti Island. He continued to make more explorations for New France in the Labrador, Hudson Bay and lower St. Lawrence regions.

1697: Jolliet was appointed royal mapmaker.

1700: Jolliet died in a Quebec province.

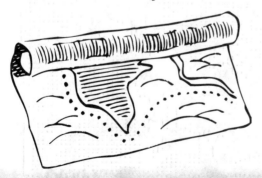

Vitus Jonassen Bering

Danish navigator who explored Russia

1681-1741

1681: Vitus Jonassen Bering was born in Horsens, Denmark.

1703: Bering entered the newly formed navy of Peter I the Great, of Russia.

1724: Because of his bravery in the wars with Sweden, Bering was appointed to lead an important expedition. He was asked to explore the water routes between Siberia and North America. He was to determine whether or not the continents of Asia and America were joined.

1728: Bering left from a port in the east of Kamchatka. He sailed northward and discovered the narrow channel that connects the Arctic Ocean and the Pacific Ocean. It was named the Bering Strait in his honor. Because of bad weather, he did not see the North American continent. But he did prove that Asia and North America were not connected.

1730: He returned to Saint Petersburg, Russia. He wanted to undertake another exploration of northeastern Siberia.

1733: Bering took command of a second larger and more ambitious venture. Eventually he was responsible for mapping large areas of the northern Siberian coast.

1741: Bering sailed from Siberia to the North American continent. He sailed into the Gulf of Alaska and landed on Kayak Island. He explored the southern coast of Alaska.

On his return voyage, Bering encountered the Aleutian Islands. Then Bering and most of his crew became ill with scurvy. His ship was wrecked on an uninhabited island that was later named after him. Bering died from the extreme cold weather on Bering Island. A few survivors built a vessel and returned to Kamchatka in 1742.

Daniel Boone

American pioneer and explorer

1734-1820

1734: On November 2, Daniel Boone was born near Reading, Pennsylvania, to a Quaker farming family. He was the sixth of eleven children. Daniel probably had no formal education, but he learned about livestock, wagons, weaving and blacksmithing. His aunt taught him to read and write.

1746: On Daniel's 12th birthday, his father gave him a new rifle. He was already a skilled hunter and trapper. He developed great strength and quickness.

1753: Boone's family moved to a place on the Yadkin River in what is now North Carolina.

1755: Daniel served with the British forces in the French and Indian War. He met John Finley, a hunter who had seen the "wild west." His stories captured Daniel's imagination. During the war, Daniel joined General Braddock's expedition that attempted to drive the French from Fort Duquesne (Pittsburgh). An ambush ended the expedition, but Boone escaped.

Boone married his childhood sweetheart, Rebeccah Bryan. She often traveled with him.

1767-1768: Boone made his first exploration in the region around the Kentucky River, into the edge of Kentucky. John Finley convinced him to go on an even greater adventure.

1769-1771: Boone made his second and most important trip, along with Finley and four others. They followed a gorge through the Cumberland Mountains (known as the Cumberland Gap) and explored eastern Kentucky. Once, Boone and a companion were captured by Native Americans. They escaped while their captors slept.

1773-1774: Boone visited Kentucky again. He built a cabin at Harrodsburg. He followed the Kentucky River to its mouth.

1775: Colonel Richard Henderson from the Transylvania Company hired Boone and 30 men to cut a trail. They were told to go 300 miles through the wilderness of the Cumberland Gap to the Kentucky River. This trail became the Wilderness Road from eastern Virginia into Kentucky. It was a road by which colonists could reach Kentucky and, hopefully, settle there. Boone's group built log cabins and started a fort at the end of the trail. They named the settlement Boonesborough. (The town is now known as Boonesboro. It is 16 miles southeast of Lexington on the Kentucky River.)

1776: A Shawnee raiding party captured Boone's 14-year-old daughter, Jemima, and two friends. Boone and some companions followed the Shawnees and rescued the girls. Boone became a captain in the Virginia militia in the American Revolution.

1778: During the American Revolution, Boonesboro was attacked repeatedly. Boone was captured and taken by Shawnees, but he escaped. He walked 160 miles in four days and reached Boonesboro in time to warn the settlers that the Shawnees were about to attack.

1780s: Sometime in the early 1780s, Boone was forced to give up rights to the land around Boonesboro because his titles to the land were not valid, and he had not paid taxes.

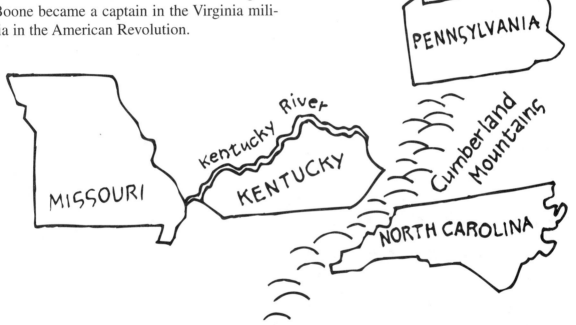

1781: Boone was elected to the Virginia legislature. The British cavalry attacked Charlottesville where the legislature was meeting. Boone was captured but was later freed.

1784: John Filson, an explorer and historian, published *The Discovery, Settlement, and Present State of Kentucky.* The book contained an "autobiography" of Daniel Boone. It helped to spread Boone's popularity as a pioneer. But Boone was still poor.

1788-1798: He left Kentucky to live near Point Pleasant, Virginia (now West Virginia). Sometimes he had a store, sold horses or guided settlers over the mountains.

1791: He was elected to the Virginia legislature a second time.

1799: Boone again moved west, near what is now St. Louis, Missouri. As he passed Cincinnati in his dugout canoe, somebody asked him why he was leaving Kentucky. Boone replied, "Too crowded." Other than two visits back to Kentucky, Boone lived the rest of his life in Missouri.

Boone received a tract of land in the Louisiana Territory west of the Mississippi River as a gift from the Spanish governor. Boone was appointed magistrate.

1803: The territory in which Boone lived became a part of the United States in the Louisiana Purchase. Since his acreage had been given to him by Spain, Boone lost the land.

1810: Boone returned to Kentucky and paid his old bills and debts. He later moved back to Missouri with his family.

1814: The U.S. Congress recognized Boone's claim to his land in Missouri.

1820: Daniel Boone died on September 26. He was buried on a hilltop overlooking the Missouri River. Later his body was moved to Kentucky.

Daniel Boone was a courageous frontiersman who kept pushing westward through the wilderness, while at the same time most Americans were content to settle along the Atlantic coast.

James Cook

British explorer
of the Pacific Ocean

1728-1779

1728: James Cook was born in Marton, Yorkshire, in England on October 27. His father was a farmer. As a youth, Cook made several voyages to the Baltic Sea.

1740: Cook was apprenticed to be a *haberdasher*, a dealer in small merchandise. Later he was apprenticed to a ship owner.

1755: Cook enlisted with the British navy.

1756-1767: He charted the waters in the North Atlantic on the coasts of Newfoundland, Nova Scotia and the Saint Lawrence River below Quebec.

1759: During the Seven Years' War, Cook was the master of the ship *Mercury*. He took part in expeditions against the French on the Saint Lawrence River. He also helped survey the river channel.

1763: After the war ended, Cook directed the schooner *Grenville* and spent four years surveying the coasts of Labrador, Newfoundland and Nova Scotia. Cook studied math in order to become a better navigator. His charts of the coasts were published because of their importance and accuracy.

1766: Cook observed a solar eclipse and used it to determine the longitude of Newfoundland. His findings were published in the Transactions of the Royal Society.

1767: Cook was commissioned as a lieutenant in the Royal Navy.

1768: Cook made his first great voyage in command of the *Endeavour*. He went in search of Tahiti in the South Pacific Ocean. His crew included an astronomer, two botanists and artists. Cook carried plenty of provisions and citrus fruits. He insisted the men use proper hygiene. For all these reasons, his crew was able to avoid the plague of scurvy.

1769: Cook reached Tahiti and, with his crew, observed the movement of the planet Venus across the sun. He reached New Zealand, which was first explored by Abel Tasman in 1642. Cook spent six months exploring New Zealand and mapping its coasts. He showed that the North Island and South Island of New Zealand were two islands and not a continent.

From New Zealand, Cook sailed to the east coast of Australia. After the *Endeavour* became grounded on a coral reef, Cook was able to save his ship. In all, Cook surveyed about 2000 miles of Australian coast. He proved the existence of a passage between Australia and New Guinea.

1771: Cook returned to England by way of the Indian Ocean and the Cape of Good Hope.

1772: Cook's first voyage had not completely disproved legends that a major southern continent existed. Therefore, Cook began another expedition in the South Pacific in search of the legendary southern continent, *Terra Australis,* of which Africa was thought to be a part. Cook's ship this time was the *Resolution.* He was also accompanied by the *Adventure.* They left Plymouth, England, in July and headed for the Cape of Good Hope in southern Africa. Then the ships traveled further south.

1773: On January 16, Cook made the first European crossing of the Antarctic Circle. But he did not sail far enough south to discover the true Antarctic continent. The *Resolution* lost touch with its partnering ship, and the *Adventure* returned to England. It became the first vessel to go around the globe from west to east. The *Resolution* again crossed the Antarctic Circle. In this same year, Cook sailed to the south Pacific islands which are now named after him.

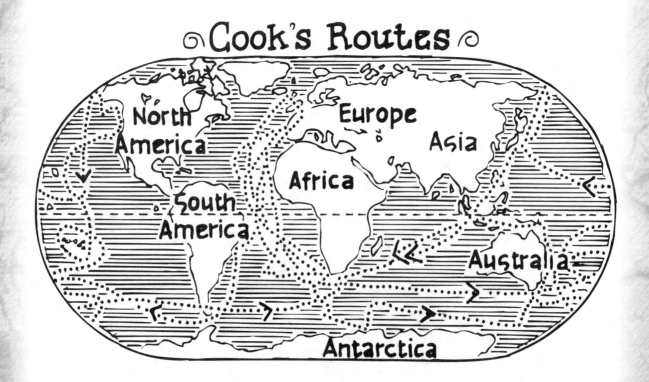

1774: Cook made charts of New Hebrides, the Marquesas and Easter Island. He explored several other Pacific Islands. The maps he made of the southern Pacific region are much like the ones we use today.

1775: Cook finally returned from his second trip in July. Only one crew member had died of disease. Cook received the Copley medal for his success in preventing scurvy by feeding his crewmates juice made from sauerkraut. Cook was made a fellow in the Royal Society for his successful expeditions. Cook had proved that there was no great continent in the temperate regions, but he was convinced that there was an icy Antarctic continent.

1776: Cook was promoted to captain and began his third voyage. His goal was to determine if there was a passage between the Atlantic and Pacific Oceans north of the North American continent (the Northwest Passage). He plannned to search for the passage from the Pacific side. At the Cape of Good Hope, Cook was joined by the *Discovery*. The two ships visited Tahiti.

1778: On the trip Captain Cook came upon some islands in the Pacific that he named the Sandwich Islands, after the Earl of Sandwich. They were later renamed the Hawaiian Islands. Sailing toward North America, Cook then made extensive explorations of the northwestern coast of North America. He charted the coastline as far north as the Bering Strait. Cook found ice but no passage.

1779: Cook returned to the Sandwich Islands where he had won the friendship of the local people. While there, one of his ship's boats was stolen. He took some of his crew ashore to recover the boat. A skirmish took place in which Cook was killed on February 14.

Captain James Cook has been called the greatest explorer of the 18th century. He surveyed and charted thousands of miles of coastline in the Pacific Ocean. He applied scientific methods to exploration and to cartography. He opened the northwest American coast to trade and colonization. He avoided scurvy. He handled his ships and crews remarkably well and conducted his expeditions in a very peaceful manner.

Louis Antoine de Bougainville

First Frenchman to sail around the world
1729-1811

1729: Louis Antoine de Bougainville was born in Paris.

1754: He abandoned his studies in law to join the French army. Also about this time, at the age of 25, he wrote a paper on integral calculus and was made a member of the Royal Society of London.

During the French and Indian War, which began in 1754, Bougainville was an aide to a marquis in Canada. He helped defend Ticonderoga and Quebec. Later, Bougainville fought Germany during the Seven Years' War.

1764: Bougainville established a French colony in the Falkland Islands. But since Spain claimed the islands, the colony had to be abandoned.

1766-1769: Bougainville became the first Frenchman to sail around the world. He visited Tahiti, the Samoan Islands, the Solomon Islands and the New Hebrides. Several naturalists and astronomers went with him. The group made many geographic and scientific discoveries.

Also on board Bougainville's ship was a young French woman named Jeanne Baret. She sailed disguised as a male servant to one of the scientists on the expedition. She was probably the first woman to sail around the world. Her true identity was not revealed until after the group had reached Tahiti.

1771-1772: Bougainville published *Account of a Voyage Around the World* to share the experiences of his expedition.

1776: During the American Revolution, Bougainville served with the French fleet.

1790: After two military promotions, Bougainville retired and devoted himself to the study of science.

1796: Bougainville was chosen a member of the Institute de France and was granted many honors by Napolean.

1811: Bougainville had contracted dysentery on his voyage around the world. While he was able to combat the disease for several years, in August it acted up and caused his death. Bougainville died in Paris on August 31.

*The largest of the Solomon Islands and a strait in the Solomon Islands are named for Bougainville. A tropical flowering vine, **Bougainvillea**, is also named for this explorer and military officer.*

Alexander Mackenzie

Scottish explorer in Canada

1764*-1820

*Sources differ on exact dates in the life of this explorer.

1764: Alexander Mackenzie was born sometime around 1764 in Scotland.

1774: Mackenzie immigrated with his family to New York City.

1779: Mackenzie moved to Montreal, Canada. He was sent to Canada after the start of the American Revolution. Alexander joined the fur-trading firm later known as the North West Company.

1787: Mackenzie was sent to what is now the province of Alberta, Canada. He established Fort Chipewyan on Lake Athabasca.

1789: Mackenzie set out from Fort Chipewyan on his first expedition to explore northwest Canada on June 3. He visited areas from the Great Slave Lake to the Arctic Ocean. He traveled on the major river that was later named after him. By September 12 he was back at Fort Chipewyan. He had traveled nearly 3000 miles by canoe in 102 days.

1793: On Mackenzie's second journey, he went up the Peace River, crossed the Rocky Mountains, and followed the Fraser River and several of its tributaries. Finally, he traveled overland to the Pacific Ocean. He reached the Pacific Ocean on July 22. Mackenzie became the first European to explore the North American continent north of Mexico on an overland journey.

1799: Mackenzie left the North West Company and joined its rival, the XY Company.

1801: Mackenzie published his journals of his excursions.

1802: Alexander Mackenzie was made a knight.

1808: Sir Alexander Mackenzie returned to Scotland and spent the rest of his life there.

1820: Historians think Mackenzie died in 1820.

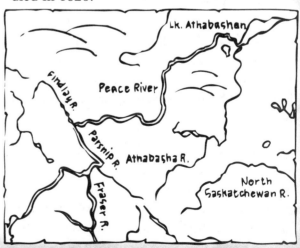

William Clark

Co-leader of the Lewis and Clark expedition

1770-1838

1770: William Clark was born in Caroline County, Virginia.

1784: The Clark family moved to the Kentucky frontier near present-day Louisville. The family established a plantation.

As a youth, Clark helped defend pioneer settlements against attacks by Native Americans. Clark followed in the footsteps of his brothers Jonathan and George Rogers Clark by joining the military.

1789: William joined a militia company under General Anthony Wayne. He was known as a courageous leader in the battles with the Native Americans in the Ohio Valley. Meriwether Lewis, who would become his co-leader exploring the Louisiana Territory, served for a short time in Clark's rifle company.

1792: Clark was commissioned as a lieutenant under General Anthony Wayne.

1794: Clark took part in the Battle of Fallen Timbers (near what is now Toledo, Ohio). This shattered the strength of the Native Americans in Ohio.

1796: Clark earned General Wayne's praise for a dangerous scouting mission. Clark also became an experienced frontier diplomat.

When Clark's family was in danger of losing their land due to debts acquired by Clark's brother George, William Clark resigned his commission and spent eight years looking after his family's interests.

1803: Lewis asked Clark to join him as co-leader on an expedition through the Louisiana Territory to the Pacific Ocean.

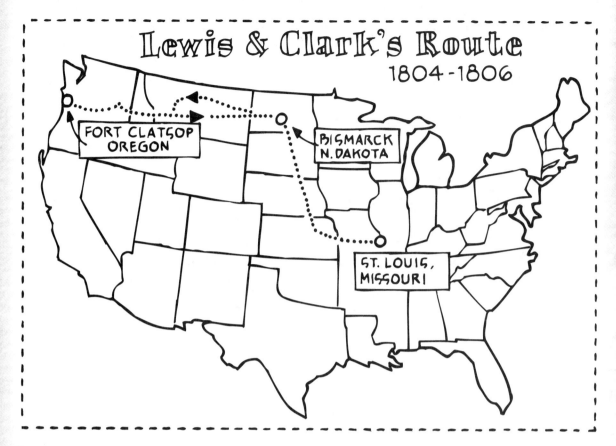

1804-1806: Lewis and Clark led the first American exploration of the territories between the Mississippi River and the Pacific Ocean. Clark was responsible for making the maps of the territories and keeping military discipline. His army experience also made him a very good negotiator and diplomat. He used those skills in many meetings with Native Americans.

The permanent party of explorers consisted of Lewis and Clark, three sergeants, 22 privates, frontiersman George Drouillard and Clark's African-American slave. They called themselves the Corps of Discovery.

The party set out on May 14, 1804. In late October they reached the villages of the Mandan Indians, where they met Sacagawea and her husband. (See biography of Sacagawea for more information.)

1805: On October 16, the Corp reached the Columbia River. On November 7, Clark wrote in his journal, "Great joy in camp. We are now in view of the ocean."

1806: In March the entire group started back east. Clark and Lewis decided to separate so they could explore more land. Clark headed for the Yellowstone River and followed it to the Missouri River, while Lewis explored a branch of the Missouri he had named the Marias.

On September 23, the expedition arrived back in St. Louis. This caused great rejoicing. They had been gone so long that many believed they were dead. Lewis and Clark both became national heroes.

1807: Clark was appointed the Superintendent of Indian Affairs. He remained in that post for three decades. He was instrumental in shaping the federal government's policies in the West. Clark was respected by Native American tribes and was known as the "red-haired chief."

1809: Lewis died and Clark became responsible for completing the report of the expedition.

1812-1815: During the War of 1812, Clark helped organize defenses in the western United States against British and Native American attacks. At the end of the war, Clark helped negotiate a series of treaties with the Native Americans.

1813-1821: Clark served as governor of the Missouri Territory.

1814: The abridged two-volume report of the Lewis and Clark expedition was published. The large map that Clark drafted for the report is a landmark in the geographic history of the West.

1820: Clark remarried when his first wife died.

1821: Clark was defeated in his run for governor of the state of Missouri. He continued to help the U.S. government with Native American issues.

1838: William Clark died in St. Louis on September 1.

In 1905, the complete set of journals from the Lewis and Clark expedition was finally published.

Meriwether Lewis

Co-leader of the Lewis and Clark expedition

1774-1809

1774: Meriwether Lewis was born to a wealthy family on a plantation near Charlottesville, Virginia. Thomas Jefferson was a neighbor and friend of the Lewis family. During his childhood, Lewis was privately tutored, and he became very interested in nature and science.

1794: Lewis joined the Virginia militia during the Whiskey Rebellion. He was soon promoted and transferred to the national army in 1795.

1795: Beginning in 1795, Meriwether Lewis served as an officer in the Northwest Territory, fighting Native Americans. In between military campaigns, Lewis lived in the wilderness to learn more about nature and Native American customs and languages.

1801: Thomas Jefferson was elected President and selected Lewis as his personal secretary. Jefferson was very interested in sending an expedition to the Pacific coast, and he charged Lewis with making preparations and doing research for this journey. Lewis estimated that $2500 would be needed, and Congress agreed to fund the trip.

1803: Lewis began to formally plan for a journey, and it became clear that the expedition would be too large for one leader. Lewis chose Captain William Clark to co-command the journey. The two spent the winter collecting supplies, training and making trips to St. Louis.

1804: Lewis, Clark and 27 men, mostly from the United States Army, headed up the Missouri River. By October, they reached the Mandan Indians, near Bismarck, North Dakota, where they met Sacagawea and her husband, Toussaint Charbonneau.

1805: On August 12, Lewis and Clark reached the top of the Continental Divide, and on November 7, the group sighted the Pacific Ocean. Throughout the trip, both Lewis and Clark kept detailed journals, which greatly aided future explorations and settlers. The two men were effective leaders and worked well together. When decisions were made, Lewis and Clark would take a vote, with Sacagawea and a freed slave, York, being counted equally with the other men.

1806: The expedition headed back from Oregon on March 23. Lewis and Clark split up to cover more territory. Lewis's party had a brief skirmish with Blackfoot Indians that left two of the Native Americans dead. In a separate incident, one of Lewis's own men shot Lewis while hunting, but Lewis fully recovered. When the group reached the Mandan Indians again, Lewis and Clark left Sacagawea and her family.

1807: Upon return from the expedition, Lewis was a national hero. President Jefferson appointed Lewis the governor of the Louisiana Territory that he had just explored. Lewis found this to be a very difficult job.

1809: Lewis left for a trip in September to Washington, D.C., to defend his work in the West. In October, Meriwether Lewis died at an inn in Tennessee.

In 1814, American diplomat Nicholas Biddle published a shortened version of Lewis's journal from the expedition. The complete version was published in 1905.

Sacagawea or Sacajawea

Guide for Lewis and Clark
1786-1812 or 1884*

*Sources differ on exact dates in the life of this explorer.

1786*: Sacagawea was born in the Snake tribe of Shoshone Indians, probably in present-day Idaho.

1800: Sacagawea was captured by the Hidatsa Indians, enemies of the Shoshones. (Sacagawea's name was made by joining the Hidatsa words for *bird* and *woman*.) She was sold as a slave to the Missouri River Mandan Indians who lived near present-day Bismarck, North Dakota. The Mandans then sold her to a Canadian trapper and French interpreter named Toussaint Charbonneau. He married Sacagawea and another Native American woman.

1803: Thomas Jefferson, President of the United States, asked Congress for $2500 to start an expedition into the Louisiana Territory. He appointed Captain Meriwether Lewis and Lieutenant William Clark to lead the expedition.

1804: Lewis and Clark traveled up the Missouri River and arrived near Bismarck, North Dakota, in November. There they met the Mandan Indians. Soon Lewis and Clark built Fort Mandan there. When they were ready to resume their expedition, they decided to hire Charbonneau and Sacagawea to help guide them. The men especially wanted Sacagawea along to help interpret and to make contact with the Shoshone Indians. (Sacagawea's brother, Cameahwait, was the chief of the Shoshones.)

1805: On February 11, Sacagawea gave birth to a boy, Jean-Baptiste Charbonneau, at Fort Mandan.

On April 7th, Lewis and Clark continued their expedition. Toussaint, Sacagawea and their baby son went with them. Sacagawea carried her baby on her back for the rest of the journey. She was the only woman to travel with the 33 men. Because a party traveling with a woman and infant would not have been viewed as threatening by Native American tribes, Sacagawea and Jean-Baptiste spared the expedition from Indian attacks.

On May 14, the boat Sacagawea was riding in nearly capsized due to high wind. She was able to save many important papers and supplies that would otherwise have been lost. The captains complimented her on her calmness during the stressful episode.

In addition to serving as interpreter, Sacagawea dug for edible plants and roots. She picked berries. These were used for food and sometimes for medicine as well.

On August 12, Captain Lewis and three men scouted 75 miles ahead of the main party. They crossed the Continental Divide. The next day, they found a group of Shoshones. Their leader, Chief Cameahwait, was Sacagawea's brother. On August 17, after five years of separation, the brother and sister were reunited. Through the interpreting chain of Sacagawea, her husband and others, the expedition was able to purchase horses it needed. Sacagawea's brother also helped the group by providing more food and men who served as guides.

On November 24, the group reached the place where the Columbia River emptied into the Pacific. A vote was held to see where the party would spend the winter. Sacagawea's vote was counted equally with the men's. The group voted to stay near present-day Astoria, Oregon, where the men built Fort Clatsop and spent the winter.

1806: On the return journey, Lewis and Clark took separate routes to learn more about the Louisiana Territory. Sacagawea traveled with Clark because he was going to the Pacific Ocean, and she very much wanted to see the great waters.

On August 14, 1806, the two groups of the expedition were reunited at Fort Mandan. Charbonneau, Sacagawea and Jean-Baptiste remained there. At the end of the trip, Sacagawea received no payment. Charbonneau was given $500.33 and 320 acres of land.

1812: Sacagawea gave birth to a daughter, Lisette. In December, one of Charbonneau's Native American wives died. It was probably Sacagawea.

1813: Eight months after the death of Sacagawea, Clark adopted Jean-Baptiste and Lisette.

1875: A woman on the Wind River Shoshone Indian Reservation in Wyoming claimed to be Sacagawea.

1884*: The second woman claiming to be Sacagawea died in 1884. Whether Sacagawea died in 1812 or 1884 has not been clearly established.

In 2000, Sacagawea was the first woman to ever be featured on a coin made by the United States mint. The golden dollar coin showed Sacagawea with Jean-Baptiste on her back.

Sacagawea is well-known for her endurance and resourcefulness. The most famous statue of Sacagawea is in Washington Park, Portland, Oregon.

David Livingstone

Scottish explorer of the Nile River in Africa

1813-1873

1813: David Livingstone was one of seven children born to poor, religious parents in Blantyre, Scotland.

1823: At the age of 10, Livingstone went to work at a cotton mill, working hard, long hours for little pay. When he was not working, Livingstone read and studied, paving the way for his academic career.

1834: David heard an appeal by British and American churches for medical missionaries to go to China. He decided that was what he wanted to do. He continued to work part-time while he studied theology and medicine. He entered Anderson College in 1836.

1838: The London Missionary Society accepted David Livingstone as a candidate to be sent to China as a missionary. But the Opium Wars in China, between the British and China, prevented Livingstone from going there. Soon, Livingstone met a missionary to Africa, Robert Moffat, who influenced Livingstone to take his work to Africa.

1840: Livingstone received a medical degree from the University of Glasgow and was ordained.

1841: Livingstone arrived in Cape Town, South Africa, on March 14. He was immediately determined to become an explorer to help open the African continent for Christianity and for western civilization. His goals were to both gain converts to Christianity and to end African slavery. Livingstone believed that if Africans could be converted to Christianity, and western free trade could be brought to Africa, slavery would be ended. Within a year, Livingstone had gone farther north into Africa than any other white man.

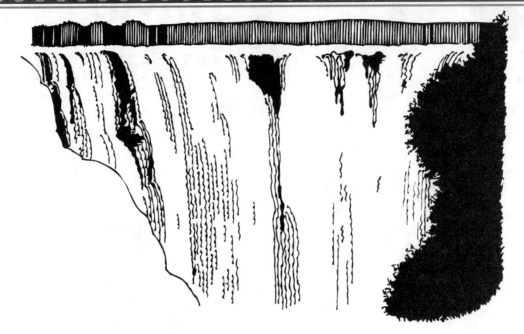

1845: David Livingstone married the daughter of Robert Moffat, Mary Moffat. For several years, Mary and their children accompanied David on his travels.

1849: Livingstone, with three other men, crossed the Kalahari Desert in southern Africa. They discovered Lake Ngami. A British exploration group awarded Livingstone a gold medal and cash prize for this feat.

1851: Livingstone again crossed the Kalahari Desert, this time with Mary and his children. On this journey, he also reached the upper Zambezi River.

1852: Livingstone returned to Cape Town in 1852. He sent his family back to England because of his wife's fragile health and his children's need for formal education.

1853: Now that his family was safe back in Scotland, David became even more adventuresome. He set out on his second major expedition in November. His goal was to reach the Atlantic Ocean and set up a new route for commerce.

1854: On May 31, Livingstone arrived at Luanda, on the Atlantic coast.

1855: Livingstone again explored the Zambezi River. He was the first European to see the waterfalls that he named the Victoria Falls, after the queen of England. He took detailed notes on the geography of the region, which greatly aided future European explorations.

1856: David Livingstone followed the Zambezi to its outlet on the Indian Ocean, becoming the first European to cross southern Africa. He then returned to England, where he published his book *Missionary Travels and Researches in South Africa* and gave many speeches. On this trip, he resigned from the London Missionary Society.

1858: He and his wife returned to Africa with an appointment from Queen Victoria to make further explorations. Livingstone spent the next five years exploring several African rivers with several assistants and ships provided by the British government. During this time, he became discouraged by the continuation of slavery in Africa.

1862: Mary, David's wife, died of malaria in April.

1864: The British government instructed Livingstone to return to England because it felt he was not making progress in his explorations. Back in England, he gave many speeches on the evils of slavery and published another book, *Narrative of an Expedition to the Zambezi and Its Tributaries.* This raised money for a venture looking for the source of the Nile River.

1866: David Livingstone explored a series of lakes and rivers to find the source of the Nile River.

1871: Livingstone became the first European to reach the Lualaba River in modern Congo, which he thought was the source of the Nile. (It is actually a source of the Congo River.) Throughout these explorations, he became physically weak and was frequently ill. Livingstone returned to Lake Tanganyika. Since 1866, Livingstone had generally been unable to send out word of his explorations. An American journalist, Henry Morton Stanley, had been sent to find Livingstone. The two met at Ujiji, and Stanley uttered the words, "Dr. Livingstone, I presume?" Together the two explored an area northeast of Lake Tanganyika.

1872: Stanley returned to England in March, but Livingstone refused to go with him.

1873: Livingstone continued exploring even as his body was weakened by dysentery. In May, he died in Chitambo, in present-day Zambia. His African companions buried his heart there before carrying his body themselves to the Indian Ocean.

1874: In April, Livingstone's remains were buried in Westminster Abbey in London. Later that year, *The Last Journals of David Livingstone* were published.

David Livingstone traveled across one-third of the African continent over 30 years. He contributed more than any other single person to opening Africa for the West. The careful observations of people and places that he shared with the British inspired countless others to journey there. His ideas of opening Africa for Christianity and commerce, and his dream of ending the slave trade, influenced many others in their explorations and colonization of Africa.

John McDouall Stuart

Scottish explorer to Australia

1815-1866

1815: John McDouall Stuart was born on September 7, in Scotland. He was the youngest of nine children. Both of John's parents died while John was young, so he and his siblings were raised by various relatives.

1839: Australia was still a new colony, and John decided to seek his fortune there after receiving his education as a civil engineer. He set foot in South Australia on January 21. He began working as a surveyor to set up official property lines in Australia.

1844: Captain Charles Sturt hired Stuart as a draftsman on his expedition into the interior of Australia. The trip took the men further into Australia than any other European had ever been. After Sturt's assistant died of scurvy, Stuart was promoted to replace him. Stuart and Sturt were stranded in the desert for six months, yet Stuart was amazed by the Australian interior. Upon their return to Adelaide on Australia's southern coast, both men had to recover from scurvy. Stuart reported that he lost the power of his limbs and was bedridden for 12 months. After his recovery, Stuart worked as a shepherd, surveyor and tutor for several years.

1858: Stuart and two other men left to explore further into South Australia. They took supplies for one month. Stuart's business partner paid for the trip. On this journey, Stuart discovered a large creek he named Chambers Creek. (Today it is known as Stuart Creek.) It was the only major discovery on this expedition, but it was a key to opening a way to the center of the continent. Stuart managed to survive and navigate his way through the bush for four months. This gained him a reputation as an excellent explorer. He was awarded a gold watch by the Royal Geographical Society.

1859: Stuart took a second journey to survey the region around Stuart Creek. He was excited about trying to be the first European to cross the continent of Australia when the government announced a monetary prize to the first person to do so.

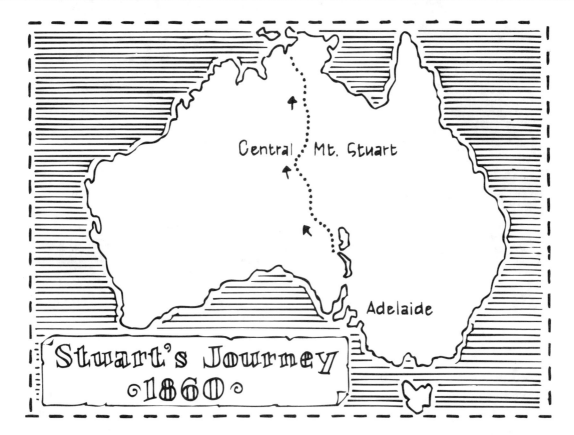

1860: Stuart and two other men left on March 2 to venture again into Australia's interior. The three became the first Europeans to enter into the Northern Territory from the south and reach the center of the continent. Stuart also discovered several rivers and mountains on this expedition before being attacked by hostile Aborigines. Stuart and his men nearly died of thirst and starvation before being acclaimed as heroes when they returned to Adelaide. The route opened by Stuart was later used to construct the Central Telegraph Line.

1861: The South Australian Parliament funded a larger expedition for Stuart to trek across Australia and designate a route for a telegraph line. The colony of Victoria funded a similar expedition led by Robert O'Hara Burke. Most of Burke's expedition died, and Stuart's first try was unsuccessful.

In October, Stuart made another journey across Australia. He reached the Indian Ocean near present-day Darwin on December 17, 1862. Finally, he had completed his dream of crossing the continent. Stuart's health problems grew more severe. When he was unable to ride his horse, he was carried on a stretcher between two horses for the rest of the journey back to Adelaide.

1865: Although he had done what many thought was impossible, John McDouall Stuart found himself without a home. His health was too poor for him to continue his work, and most of his friends in Australia had died. He returned to England in April.

1866: Stuart died on June 5, and was buried in London.

Alexandrine Pieternella Françoise Tinné

Dutch explorer of the
Nile River and North Africa

1835-1869*

*Sources differ on exact dates in the life of this explorer.

1835: On October 17, Alexandrine Pieternella Françoise Tinné was born in The Hague, Netherlands, to a wealthy family. Her father was a merchant, and her mother was a baroness. Alexandrine was tutored at home and did well in piano and photography. It is possible that a broken romance prompted her to leave home and begin an expedition.

Alexandrine's family traveled widely across Europe. She sometimes traveled alone and spent many summers in England and France. After her father's death, Alexandrine and her mother continued to travel.

1854: When she was 19, Alexandrine and her mother visited Egypt where they traveled by camel and donkey to the Red Sea. They also journeyed to the Middle East.

1857: At the age of 22, Alexandrine set out to explore the Nile River with her mother. They were accompanied by an Arab crew, servants, soldiers and animals. The group reached Wadi Haifa, Sudan, but had to stop when they encountered a large waterfall.

1862-1863: Tinné hired a small fleet of boats in Cairo, Egypt. Her mother, her aunt, several scientists, assistants and servants all went with Tinné. The group wanted to explore the Sudan region and hopefully find the source of the Nile River. They went up the Nile River as far as Gondokoro, Sudan. Beyond that, the river was not passable. No European, and especially no European woman, had ever explored any further up until then.

Alexandrine had planned to meet British explorer John Hanning Speke, who was exploring the Nile to the south. His expedition did not arrive when expected, so Tinné set off on her own to find the source of the Nile. She traveled by land and trekked into the region between the Congo and Nile Rivers, in the northeastern part of present-day Democratic Republic of the Congo. She went into regions of central Africa that were not yet mapped. She returned to Gondokoro in September. She again failed to meet up with Speke, so she traveled back to Cairo. Her mother, her aunt and two of the scientists died of fever during the trip. Tinné stayed in Cairo for several years.

1867: Tinné moved to Algiers, Algeria.

1869: Tinné tried to become the first European woman to cross the Sahara Desert. She began at Tripoli, on the Mediterranean coast of Libya. She traveled south to the city of Murzuk, at an oasis. She waited there for an Arab caravan because she wanted to continue her journey southward with them. She took a side trip to visit the nomadic Tuareg tribes. On her way to their camp, Alexandrine Pieternella Françoise Tinné was robbed and killed by her guides.

Tinné's work was valuable because she shared knowledge about geography and science in the regions she explored. She collected materials about the geology, plants, animals and climate of the upper Nile.

The few papers that remain from Tinné's travels are stored at the Royal Archives in The Hague. There is a small marker near Juba in Sudan honoring the great 19th century explorers of the Nile. The list includes the name of Alexandrine Pieternella Françoise Tinné.

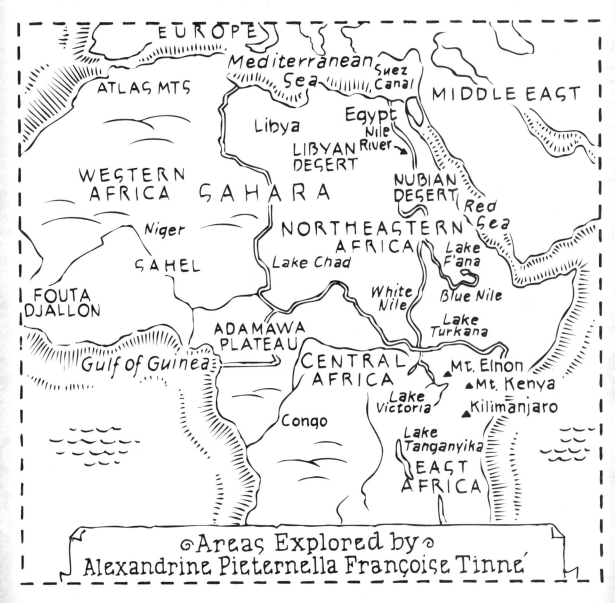

Areas Explored by Alexandrine Pieternella Françoise Tinné

Henry Morton Stanley

Welsh explorer in Africa

1841-1904

1841: Henry Morton Stanley was born in Denbigshine, Wales, on January 28. He was given his father's name, John Rowlands.

1859: After a childhood spent in extreme poverty, he ran away to sea. He landed in New Orleans, Louisiana, where he was adopted by a merchant named Henry Morton Stanley who helped him find a job. John Rowland took Stanley's name as his own.

1861: Stanley fought with the Confederate Army in the Civil War.

1862: Stanley was captured at the Battle of Shiloh. He was released when he agreed to join a federal artillery regiment, but when he came down with dysentery.

When Stanley recovered, he joined the U.S. Navy. He traveled to the Rocky Mountains and began descriptive writing.

1866: Stanley began working as a journalist for the *Missouri Democrat*. He traveled with the U.S. cavalry on operations against Native Americans in Kansas and Missouri.

1867: Stanley landed a job with the *New York Herald*. He traveled on a British campaign against the Ethiopian emperor, Theodore II.

1868: Stanley was the first to pass on the news of the fall of Magdala, Theodore's capital.

1869-1871: The *New York Herald* sent Stanley to cover the opening of the Suez Canal in Egypt, and then to report other news from Persia and India. His final assignment was to try to find David Livingstone, the Scottish explorer and missionary. Livingstone had been out of touch with the world for several years as he explored regions of Central Africa.

1871: Stanley set out eastward from Zanzibar, in charge of 2000 men. He traveled toward the place where Livingstone was suspected to be. On the way, Stanley cruelly killed native Africans because he thought that was necessary for his success However, this cruelty harmed his reputation. On November 10, Stanley found the ailing Livingstone at Ujiji, a town on Lake Tanganyika. Supposedly he said, "Dr. Livingstone, I presume?" Stanley nursed Livingstone back to health and gave him fresh supplies. He went with Livingstone on an exploration of the northern end of Lake Tanganyika.

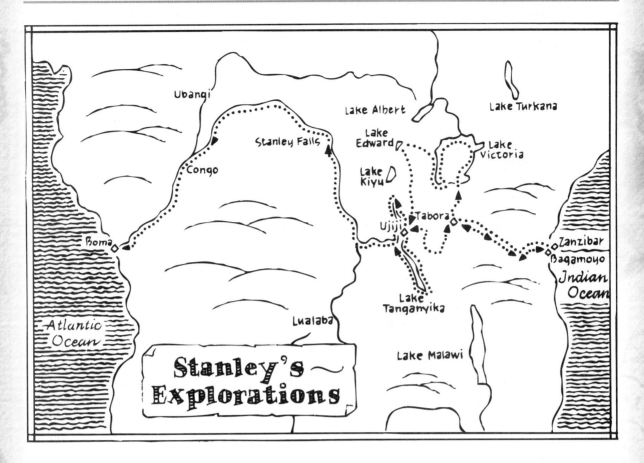

1872: Stanley published the book, *How I Found Livingstone*. It became very popular in Britain.

1873: Livingstone died, and Stanley vowed to finish the exploration begun by Livingstone. The same year, the *New York Herald* sent Stanley to report on a British military campaign in present-day Ghana.

1874: Stanley wrote about his adventures in Ethiopia and Ghana in a book called *Coomassie and Magdala: Two British Campaigns*.

1874-1877: Stanley's next expedition was financed by the *New York Herald* and the *London Daily Telegraph*. It was one of the most difficult expeditions ever assumed by a European explorer in Africa. Stanley traveled from east to west across the continent near the equator. He took 359 men with him. He fought with the natives several times along the journey. Again, Stanley was brutal.

Stanley led his men a distance of nearly 2000 miles, down the Lualaba and Congo rivers to the Atlantic Ocean, through forests and along uncharted waters. On the journey, Stanley and his men experienced disease, drowning and attacks. Once they were ambushed by cannibals. Only 108 of Stanley's original 359 men made it to the Atlantic.

This expedition led to the formation of the Congo Free State, now the Democratic Republic of the Congo. Within a few years, the nations of western Europe were competing to establish African colonies. Also as a result of this important expedition, Stanley later asked missionaries to bring Christianity to the kingdom.

1878: Stanley wrote about his adventure through Africa to the Atlantic in *Through the Dark Continent.* In it, Stanley explained the size and drainage of Lakes Victoria and Tanganyika. He proved the existence of a waterway through central Africa. Leopold, king of Belguim, paid attention to Stanley's work. He was eager to tap into Africa's riches.

1879: The British were less interested in colonizing central Africa than was Belguim. So Stanley returned to the Congo under the sponsorship of Leopold. For five years, Stanley worked to build a road and open the lower Congo for trade. He was given the African nickname *Bula Matari,* or "breaker of rocks" because of his brutality.

1884-1885: The Berlin West Africa Conference was held in which major countries worked out their claims on land in Africa. Leopold obtained rights to most of the Congo basin.

1885: Stanley became a naturalized citizen of the United States.

1888-1889: Stanley led a journey to reach Emin Pasha, a German explorer in Africa. Pasha was a medical officer in the Sudan who had been cut off by an uprising led by a holy man known as the Mahdi. Stanley traveled through lands no European had visited.

1890: After Stanley's journey to rescue Pasha, he settled down and married. He went on speaking tours in the United States, Australia and New Zealand until 1892. Then he returned to England and became a British citizen again.

1895: Stanley won a seat in the British parliament. He held it until 1900.

1897: Stanley made his last trip to Africa. He visited land held by Britain in southern Africa. He wrote *Through South Africa.*

1899: He was knighted by Queen Victoria.

1904: Stanley died on May 10. He was buried in Westminster Abbey, London. By this time, nearly all of Africa belonged to Europe.

Robert Edwin Peary

American explorer, first to reach the North Pole

1856-1920

1856: Robert Edwin Peary was born in Cresson, Pennsylvania. He was educated at Bowdoin College. He became a land surveyor and draftsman.

1881: At the age of 25, Peary became a civil engineer in the U.S. Navy.

1886: Peary visited the Greenland ice cap on his first trip to the Far North. The lure of Arctic exploration was in his blood. This was the first of seven polar expeditions for him.

1887: Peary was chief engineer of the Nicaragua canal survey. He hired African-American Matthew Hensen to work with him.

1888: Peary married Josephine Diebitsch. She shared his ambition for exploration and spent several winters with him in the Arctic.

1891-1892: Peary crossed Greenland to the Arctic Ocean and proved it was an island. He explored and mapped the northern coastline of Greenland. Traveling with Peary was his wife, Henson and Dr. Frederick Cook, a surgeon.

On July 4, 1892, Peary discovered and named Independence Bay on the northern coast of Greenland.

1893: Robert and Josephine's first child was born in Greenland.

1893-1895: Peary made his third voyage to try to reach the North Pole.

1898-1902: Peary made his fourth attempt to get to the North Pole.

1905-1906: In this effort to reach his goal, Peary reached a location just 174 miles away from the Pole.

1908: Peary set off on a sledge expedition to reach the North Pole. He traveled with 59 Inuits and seven others.

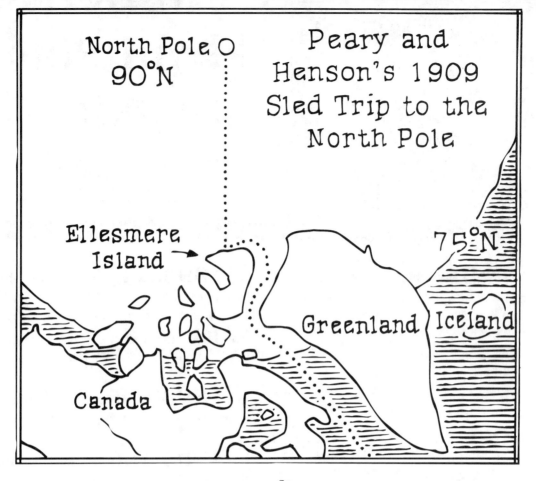

1909: On April 6, Robert Peary and part of his original team reached the North Pole. Then it took five months for them to return to Greenland and make the announcement. Only Henson and four Inuits were with Peary on the final leg of the expedition. It was a 950-mile round trip. It took the men and their dog teams more than 50 days. Henson planted the American flag on the North Pole.

(On September 1, Dr. Frederick Cook announced that he had reached the North Pole in the previous year. His claims were generally not taken seriously.)

On September 6, when word finally reached the rest of the world of Peary's accomplishment, this message was flashed around the world by cable and telegraph: "Stars and Stripes nailed to the North Pole—Peary."

1911: The U.S. Congress retired Peary with the rank of rear admiral.

1920: Peary died on February 20, in Washington, D.C.

In 1940, Frederick Cook died. He claimed until his death that he had beaten Peary to the North Pole.

In 1989, a new study conducted by the National Geographic Society supported Peary's claims that he was first to reach the North Pole.

Mary Henrietta Kingsley

British author and explorer in Africa

1862-1900

1862: Mary Henrietta Kingsley was born in London, England, on October 13.

Her father was a doctor who traveled extensively. She had no formal education, but she read regularly from her father's scientific library. Mary cared for her parents when they became invalids.

1893: After her parents died, Mary decided to study native religion and law in West Africa. She sailed to the coast of Nigeria, and then traveled inland. From the Niger River region in the north, she traveled southward as far as the region now known as northern Angola.

1894: Kinglsey returned to England briefly before revisiting West Africa. She stopped along the coast of what are now Cameroon and Gabon. From Gabon, she traveled by steamboat up the Ogooue River. Then she journeyed by canoe into the Great Forest Region, an area rarely visited by Europeans. She studied the culture of the Fang people and then returned to the Cameroon coast.

1895: In July, Kingsley explored the lower reaches of the Ogooue River. In September, she climbed the Cameroon Mountains, the area's highest peak. She returned to England later that year.

1897: Kingsley's first book, *Travels in West Africa,* was published.

1899: Kingsley's later books, *West African Studies* and *The Story of West Africa,* were published. Kingsley was considered an expert in African culture, and her writings influenced anthropologists, historians and policymakers for many years. She supported British trade interests in Africa, but she also asserted that native African cultures should be respected.

Kingsley made her final journey to Africa in 1899. She planned to visit West Africa again, but the Boer War broke out there. She went to South Africa instead.

1900: She worked in Cape Town as a nurse, caring for prisoners of war from the Boer War. She contracted typhoid fever. Mary Henrietta Kingsley died in Cape Town, South Africa, on June 3.

Robert Falcon Scott

English explorer of Antarctica

1868-1912

1868: Robert Falcon Scott was born in Devonport, England.

1880: At age 12, Robert entered the royal navy as a cadet.

1897: He became a first lieutenant in the British navy.

1900: Scott was chosen to lead the 1901-1904 National Antarctic Expedition of the *Discovery*. On his expedition he conducted scientific experiments.

1901: Scott launched his first journey to Antarctica. He established a base at McMurdo Sound in Antarctica. His team explored new areas of the continent. They traveled east of the Ross Ice Shelf and sledged over Victoria Land. Scott set a record for traveling the farthest south, past 81° latitude.

1904: The expedition returned to England. Scott's team had made many important scientific discoveries on the journey.

1905: He wrote of his trip and entitled it *The Voyage of the Discovery*.

1908: Scott married Kathleen Bruce. They had one son.

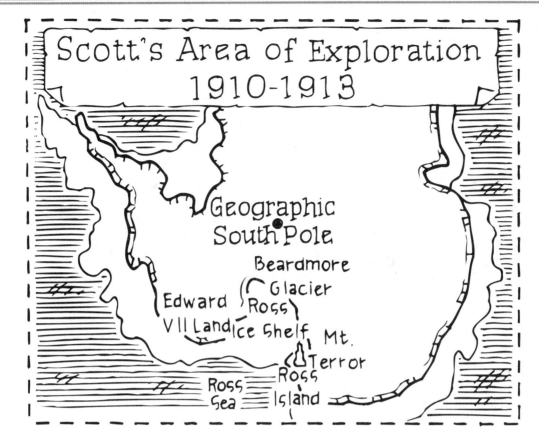

1910: Scott took a team of men and returned to Antarctica. Scott intended to be the first man to see the South Pole.

1911: Scott's team landed at their base at McMurdo Sound. His motor sledge journey started with 11 men. By December 31, seven of the men had to go back to the base camp due to broken equipment and severe weather.

Scott and four others (Captain Lawrence Oates, Petty Officer Edgar Evans, Henry R. Bowers and Dr. Edward Wilson) completed the trip of 1842 miles, which was the longest continuous sledge journey ever made in the polar regions.

1912: Scott and his four companions reached the South Pole on January 18. They found a Norwegian tent and flag already standing there. The explorer Roald Amundsen had arrived there five weeks earlier, in December 1911.

On their 800-mile return to the base, Evans died from a fall. Oates walked out into the cold to save the others but died in a blizzard. In March, Scott, Bowers and Wilson died of starvation and exposure.

Their bodies were found by a search party in November. They had come within 11 miles of a station where they had stored food and fuel supplies. A snow monument with a cross was placed at the site.

In 1913, Robert Falcon Scott's diaries and documents were published as **Scott's Last Expedition.** *His last diary ended with these words: "We shall stick it out to the end, but we are getting weaker, of course, and the end cannot be far. It seems a pity, but I do not think I can write any more."*

Roald Engelbregt Grauning Amundsen

Norwegian explorer of the South Pole

1872-1928

1872: Roald (pronounced **ROH-ahl**) Amundsen was born in Borge, Norway, near Oslo. His father was a shipowner. He died when Roald was just 14 years old. In school Roald read stories about polar explorers and decided he wanted to become one too.

1895: Although Roald wanted to be an explorer, he studied medicine for two years at the University of Christiana (now the University of Oslo) because his mother wanted him to become a doctor. He also entered the Norwegian navy at some point.

1897: Amundsen left medical school after his mother died. He went on a Belgian expedition to the Antarctic. He was the first mate of the ship *Belgica*. When he returned to Norway, he prepared to make his own journey.

1903: Amundsen set sail in the ship *Gjöa*, hoping to locate the magnetic North Pole. He located it and then stayed in Greenland for 19 months studying it. After analyzing the magnetic North Pole, he found it has no stationary position.

1905: While on his voyage to study the magnetic North Pole, Amundsen also traveled the complete Northwest Passage, from the Atlantic to the Pacific. He went through bays, straits and sounds north of Canada. Finally, he accomplished what explorers had been trying to do for more than 300 years.

1906: He returned from his trip to Greenland.

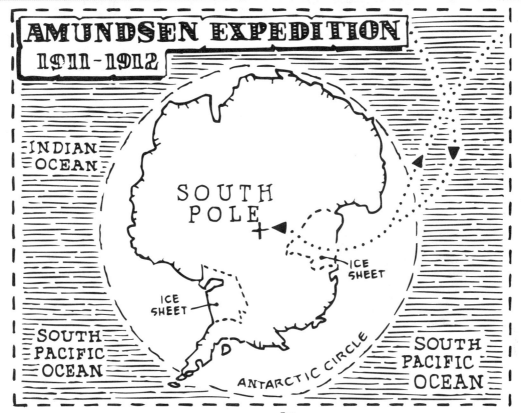

1910: Amundsen planned next to drift across the North Pole. But after learning that Robert Peary, an American, had reached the North Pole in 1909, Amundsen set sail for the South Pole. He gained fame as a successful Antarctic explorer. He lived in Antarctica for more than a year and conducted many scientific experiments.

1911: On December 14, Amundsen became the first person to reach the South Pole. He knew the British explorer, Robert Scott, was heading there also. Amundsen left Scott a sympathetic note at the pole, which Scott found when he got to the pole 35 days later.

Amundsen took a shorter route than Scott. He used dogs to pull the sleds and to provide for food on the return trip. He had good weather conditions. But some have said his success was due mostly to the fact that he had great knowledge of polar conditions, he paid great attention to details and he had great physical strength and endurance.

1912: Amundsen returned to Norway, but his exploration plans were interrupted by the beginning of World War I.

1918: Amundsen sailed from Norway to the Arctic regions in the ship *Maud*. Amundsen tried to drift across the North Pole from Asia to North America with the ice currents. But the currents were unpredictable and the ship was unable to go through the solid ice. He was forced to take a more southerly route.

1920: Amundsen reached Nome, Alaska, by way of the Northeast Passage, the Siberian waters that connect the Atlantic with the Pacific.

1922: Amundsen attempted to reach the North Pole again by ship and airplane. These attempts were also unsuccessful.

1924: He came to the United States in order to obtain funding for more exploration.

1926: Roald Amundsen succeeded in flying over the North Pole in the dirigible *Norge*. The flight went from Spitsbergen, Norway, to Teller, Alaska—2700 miles. It took more than 70 hours. Umberto Nobile, an Italian, was the pilot for the flight. The dirigible had been designed and built by Nobile and paid for by the Italian government. Amundsen and Nobile argued over which country should get the credit for the flight.

1928: In June, Nobile took a second dirigible flight over the Arctic and crashed. Amundsen came out of retirement to try to rescue him. Nobile was rescued, but Amundsen's plane crashed. The remains of the plane were found months later near Tromso, Norway.

Roald Amundsen was the first man to reach the South Pole, the first to sail around the world through the Northwest and Northeast Passages and the first to fly over the North Pole in a dirigible. Amundsen was a famous speaker and magazine writer. He wrote several books including **The South Pole** *(1912) and co-wrote* **First Crossing of the Polar Sea** *(1927) and* **My Life as an Explorer** *(1927).*

Roy Chapman Andrews

American explorer of the Gobi Desert

1884-1960

1884: Roy Chapman Andrews was born on January 26 in Beloit, Wisconsin. As a boy, he spent his time playing in the woods and fields near his house.

1902: Andrews entered Beloit College. He paid for his college education by teaching himself taxidermy and practicing it part time. Andrews majored in English but also took classes on subjects that were then rarely studied, such as archaeology.

1906: Andrews began working at the American Museum of Natural History in New York City. At first, he swept floors and helped with taxidermy procedures. Eventually he was promoted. He began making whaling expeditions to Korea and Alaska to collect samples for the museum. He built one of the world's best collections of sea mammals.

1920: Andrews proposed a trip to the Gobi Desert in Mongolia to search for fossils. Despite much criticism and dangers such as bandits, sandstorms, snakes and extreme desert temperatures, the museum president, Henry Fairfield Osborn, supported Andrews' idea.

1922: Andrews began his Mongolian expedition in April with a caravan of Dodge cars and camels to transport food, scientists and explorers, and other supplies across the desert. The first major find of the trip were remains of a *Baluchitherium*, the largest land mammal known to have ever lived.

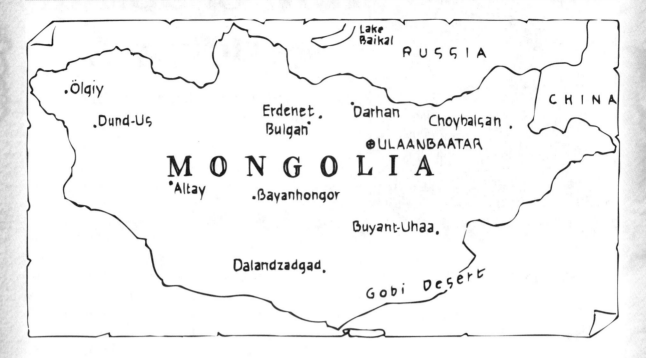

1923: After spending the winter in China, Andrews and his team returned to the Gobi Desert and made their most famous find on July 13. An assistant on the expedition, George Olson, made the first human discovery of dinosaur eggs. During the freezing desert nights, members of the expedition frequently dealt with dozens of snakes seeking warmth in their tents.

1925: A skull found during 1923 was proven to be from a mammal that co-existed with dinosaurs, the first discovery of its kind.

1926: Andrews' party was for the first time threatened by the civil unrest in the region. Soldiers fired upon his caravan for three miles of travel. His party was unharmed, but it became clear that Mongolia was no longer safe for exploration. Andrews made another trek into Mongolia in 1928. His findings were seized by authorities. Andrews spent six weeks in negotiations to have the fossils returned to him.

1930: Andrews made one last venture into Mongolia. His team found a group of *mastodons*, or large-tusked mammals.

1935: Andrews was appointed director of the American Museum of Natural History.

1942: Andrews retired to write about his explorations and discoveries. He published several books, including *The New Conquest of Central Asia* (1932); *This Amazing Planet* (1940); and an autobiography, *Under a Lucky Star* (1943). Andrews and his adventures were also featured in several movies.

1960: Andrews died on March 11 in Carmel, California.

In 1990, the American Museum of Natural History was invited by the Mongolian government to return to the Gobi Desert for further expeditions. Scientists began by looking for fossils that Andrews discovered 60 years earlier.

Richard E. Byrd

American polar explorer

1888-1957

1888: Richard Evelyn Byrd was born in Winchester, Virginia. His father was a lawyer. His brother, Harry Flood Byrd, became a U.S. senator.

1900: When Richard was 12, his parents let him make a trip around the world alone.

1912: Byrd graduated from the U.S. Naval Academy.

1916: Byrd was retired from the Navy because of an injured ankle. Within a few months, he was reassigned to a Navy flying school.

1918: During the final months of World War I, Byrd commanded an air station in Nova Scotia.

1926: Byrd's first polar expedition was to the Arctic. He navigated the first plane to fly over the North Pole on May 9. Floyd Bennett was his pilot. Byrd was awarded the Medal of Honor for this achievement. (In 1996, an investigation concluded that Byrd and Bennett never made it to the pole.)

1927: A few weeks after Charles Lindbergh's solo flight across the Atlantic, Byrd crossed the ocean with a crew of three in the three-motored plane, *America*. His flight lasted 42 hours and ended in a crash landing along the coast of France.

1928-1930: Byrd made his first expedition to the Antarctic. He claimed a large territory for the U.S. He named it Marie Byrd Land, after his wife. He established a base called Little America on the Bay of Whales.

In 1929, with the help of pilot Bernt Balchen, Byrd made the first airplane flight over the South Pole. Byrd was made rear admiral in 1930.

1933-1935: Byrd claimed more land for the United States. He spent five months alone in a hut, about 123 miles south of his base so that he could study inland weather conditions in the Antarctic. He survived temperatures as low as –76° Fahrenheit. Byrd became ill from fumes caused by a clogged chimney but refused to call for help. He was rescued in early spring.

1939-41: Byrd went on his third trip to Antarctica. He discovered the southern limit of the Pacific Ocean.

1946-1947: Byrd commanded Operation High Jump, a project by the Navy to discover and map large areas of Antarctica.

1955: Byrd was named head of Operation Deep Freeze, an Antarctic expedition organized by the U.S. government.

1956: Byrd took his last flight over the South Pole.

1957: On March 11, Byrd died at his home in Boston. He was hailed as an international hero.

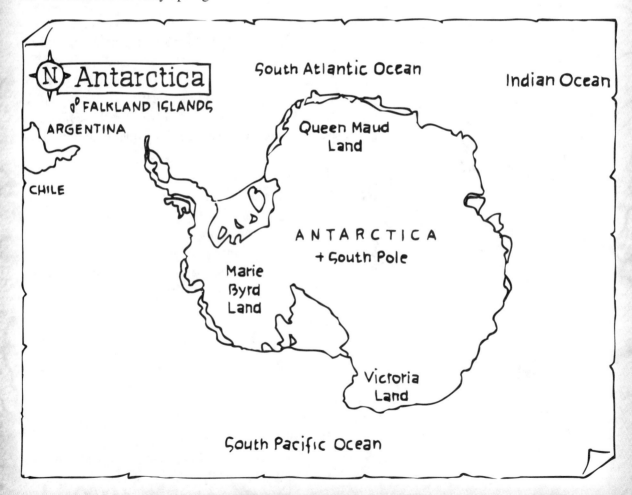

Sir Edmund Hillary

First to reach the summit of Mount Everest

1919-

1919: Edmund Percival Hillary was born in Auckland, New Zealand, on July 20. His father was a beekeeper and taught Edmund the trade. Young Edmund read constantly during the long rides to school. He dreamed about exploring Antarctica.

Edmund started mountain climbing as a teenager in the Southern Alps of New Zealand.

He served for two years in New Zealand Air Force as a navigator during World War II, but was discharged after an accident.

1924: George Leigh-Mallory, an accomplished mountaineer, died trying to summit Mount Everest.

1951-1952: Hillary joined expeditions from New Zealand to the Himalayas that verified possible routes to the summit of Mount Everest (29,028 feet). Hillary had prepared for these expeditions by climbing extensively in Austrian and Swiss Alps and climbing 11 peaks taller than 20,000 feet in the Himalayan Mountains.

1953: British army officer Sir John Hunt organized an attempt to summit Mount Everest with more than 400 climbers. On May 29, Hillary and Tensing Norkay, a Nepalese mountain climber, reached the summit of the highest mountain in the world, Mount Everest. Their accomplishment was announced the night before the coronation of Queen Elizabeth II. Hillary went to Britain and was knighted by Queen Elizabeth II. Norkay was given Great Britain's highest civilian award, the George Medal. Hillary became an instant celebrity and used his fame to raise funds for Nepalese building projects. (Norkay had been involved in five previous attempted climbs of Everest.)

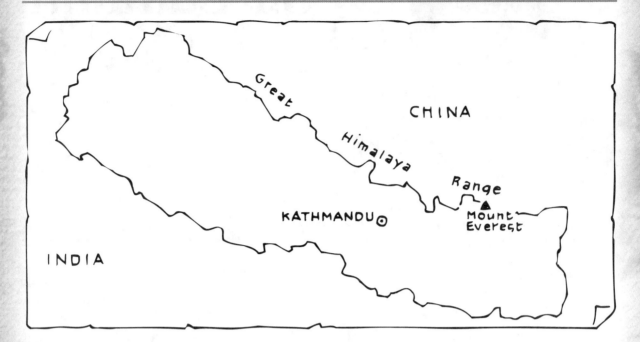

1958: The New Zealand group of the British Commonwealth Transantarctic Expedition, led by Hillary, ventured across Antarctica on tractors and became the first people to reach the South Pole by traveling overland since Robert F. Scott.

1961: Hillary established the Himalayan Trust, an organization dedicated to improving the lives of Nepalese citizens by building schools, medical clinics, hospitals and airstrips. Hillary then became involved in the effort to protect the environment around Everest by declaring the land a national park and passing laws regulating deforestation. He persuaded the New Zealand government to finance these efforts. Throughout the 1960s, Hillary made several more expeditions into the Himalayan Mountains.

1967: On another expedition to Antarctica, Hillary scaled Mount Herschel (10,941 feet). This was the first time the feat had ever been accomplished.

1975: *Nothing Venture, Nothing Win*, Hillary's autobiography and one of his several books, was published. In the greatest tragedy of Hillary's life, his wife and daughter died in a crash on a trip to Nepal. (Hillary later remarried.)

1977: In September and October, Hillary journeyed by jet boat from the mouth of the Ganges River to its source in the Himalayan Mountains.

1985: Hillary was appointed the High Commissioner to India, Nepal and Bangladesh. Based in New Delhi, India, he spent the next four years continuing his fundraising efforts and improving relations between his country and the region. Also that year, Hillary and Neil Armstrong landed at the North Pole in a small plane, making Hillary the first man to explore both poles and the summit of the world's tallest mountain.

(At the time of this writing, Hillary was still alive.)

Additional Explorers for Further Study

This book does not allow sufficient space for the biographies of *every* significant explorer. Listed here are a few more adventurers for your research. Some are rather obscure and will require extra effort, but all are very interesting. Use the outline on page 109 to record your findings.

Pedro Álvares Cabral (1460-1526*) Portuguese explorer in South America, first European to visit Brazil

Sebastian Cabot (1476-1557*) Italian-born explorer and cartographer, made expeditions to North and South America for Spain and Britain

Giovanni da Verrazzano or Verrazano (1480-1527*) Italian explorer for the French in North America, first European to sail into New York Bay

Francisco de Orellana (1490-1549*) Spanish explorer of Amazon River

Estavanico or Estaban (1500-1539*) slave from Morocco, one of the first explorers of the southwestern United States

Martin Frobisher (1535-1594*) British navigator, first to seek the Northwest Passage

John Davis or Davys (1550-1605*) British navigator, searched for Northwest Passage

Pierre Gaultier de Varennes (1685-1749) Canadian explorer in western Canada and the United States

Willem Barents or Barentz (1550-1597*) Dutch navigator in Arctic Ocean, made three voyages in the 1590s in search of the Northeast Passage

Willem Jansz or Janzsoon (1570 - 1630*) Dutch navigator thought to be the first European to land on the Australian continent

*Sources differ on exact dates in the life of this explorer.

⊛**George Vancouver** (1757-1798) British explorer of Pacific Coast of North America, sailed with Captain James Cook

⊛**Zebulon Montgomery Pike** (1779–1813) U.S. Army officer, explorer of Rocky Mountains

⊛**Ida Pfeiffer** (1797-1858) Austrian explorer and travel writer

⊛**John Charles Frémont** (1813-1890) U.S. explorer, army officer and politician who mapped much of the far western United States, including the Oregon Trail

⊛**Isabella Lucy Bird Bishop** (1831-1904) British adventurer, world traveler and writer who circled the globe three times, according to some sources

⊛**Nils Nordenskjöld** (1832-1901) Swedish explorer, first to pass through Northeast Passage

⊛**Nikolai Przhevlaski** (1839-1888) Russian explorer in Mongolia and western China

⊛**May French Sheldon** (1847-1936) American explorer in Africa, inspired by Henry Morton Stanley; one of the first white women to visit parts of eastern and central Africa

⊛**Fridtjof Nansen** (1861-1930) Norwegian explorer to the Arctic, also a statesman who won the Nobel Peace Prize in 1922

⊛**Gertrude Bell** (1868-1926) British traveler and writer in Orient, helped British armies find routes in Arabia during World War I

⊛**Donald Baxter MacMillan** (1874-1970) U.S. explorer to the Arctic with Robert Peary, authored books on Arctic exploration

⊛**Louise Arner Boyd** (1887-1972) U.S. explorer, first woman to fly over North Pole

⊛**Vivian Ernest Fuchs** (1908-1999) English explorer in Antarctica and geologist

⊛**Tensing Norkay** (1914-1986) Nepalese guide, traveled with Sir Edmund Hillary as first to reach summit of Mount Everest

Information on Additional Explorers

Name of explorer: _____

Years of birth and death: _____

Interesting events in the life of this explorer, along with the years:

_____

_____

_____

_____

_____

_____

_____

_____

_____

_____

_____

_____

_____

Bibliography

Here is a list of many of the resources used in the research of this book. Additional Web sites were also accessed to find specific information on individual explorers.

Compton's Encyclopedia 2000. CD-ROM. The Learning Company, 1999.

Dorling Kindersley Children's Illustrated Encyclopedia. Fifth ed. New York, NY: DK, Ltd., 2000.

Encarta 99. CD-ROM. Microsoft, 1999.

Grosseck, Joyce C. and Elizabeth Attwood. *Great Explorers.* Grand Rapids, MI: Gateway Press, Inc., 1988.

Magnusson, Magnus, ed. *Chambers Biographical Dictionary.* Fifth ed., Edinburgh: W & R Chambers Ltd., 1995.

Matthews, Rupert. *Explorer, Eyewitness Books.* New York, NY: DK, Inc., 2005.

The Software Toolworks. CD-ROM. Grolier, Inc., 1992.

Various. Enchanted Learning Software. http://www.enchantedlearning.com/explorers/

Various. Encyclopedia Britannica. http://britannica.com

Various. Public Broadcasting Service. http://pbs.org

Various. Worldbook Encyclopedia and Learning Resources. http://worldbook.com

Webster's New Universal Encyclopedia. Helicon, Ltd., 1997.